ENGINEERING

COMMUNICATIONS

PRENTICE-HALL SERIES IN ENGINEERING DESIGN

JAMES B. RESWICK, *editor*

FUNDAMENTALS OF ENGINEERING DESIGN

Creative Synthesis in Design—*Alger and Hays*
Introduction to Design—*Asimow*
Engineering Communications—*Rosenstein, Rathbone, and Schneerer*
Product Design and Decision Theory—*Starr*

ENGINEERING DESIGN SERIES—BOOKS NOW IN PREPARATION

Fundamentals of Engineering Design

Reliability in Engineering Design—
  Reethof and Queen

Studies in Engineering Design

Appliance Design—*Woodson*
Modern Gear Dynamics—*Richardson*
Shock and Vibration—*Crede*

PRENTICE-HALL INTERNATIONAL, INC. *London*
PRENTICE-HALL OF AUSTRALIA, PTY., LTD. *Sydney*
PRENTICE-HALL OF CANADA, LTD. *Toronto*
PRENTICE-HALL OF INDIA (PRIVATE) LTD. *New Delhi*
PRENTICE-HALL OF JAPAN, INC. *Tokyo*
PRENTICE-HALL DE MEXICO, S. A. *Mexico City*

ENGINEERING COMMUNICATIONS

ALLEN B. ROSENSTEIN

Associate Professor of Engineering
University of California, Los Angeles

ROBERT R. RATHBONE

Associate Professor of English
Massachusetts Institute of Technology

WILLIAM F. SCHNEERER

Associate Professor of Engineering Graphics
Case Institute of Technology

PRENTICE-HALL, INC., Englewood Cliffs, N. J.

© 1964 by Prentice-Hall, Inc., Englewood Cliffs, N. J. All rights reserved.
No part of this book may be reproduced, by mimeograph or any other means, without permission in writing from the publisher.

Library of Congress Catalog Card Number 64-00000

Printed in the United States of America

C27726P, C27727C

FOREWORD

All engineering is dependent upon the accurate, economical, and rapid transmission and processing of information. Communication is an indispensable activity in the engineering design process. The engineer must communicate with himself and he must communicate with others. The theory and practice of communication are the substance of the chapters which comprise this book.

A unified approach to engineering communication is presented. The basic concepts of information theory are carefully developed and shown to be an integral part of the modern theory of engineering design. The fundamentals of efficient engineering communications are presented from the viewpoint of information theory. This provides the logical basis to which the rules for effective reading, writing, and drawing are related.

In chapters one through three Professor Rosenstein presents a framework for communication based on information theory concepts. Everyday experiences are used to illustrate the meaning and application of the fundamentals of information theory. Professor Rosenstein's purpose is twofold: (1) to introduce the student to the quantitative aspect of information theory and (2) to provide him with a rational theory that will enable him to communicate more effectively. With an understanding of the basic requirements for all engineering communication, the reader will consequently be able to evaluate new situations for which he does not possess rote rules of behavior.

The practice of communication is succinctly developed by Professors Rathbone and Schneerer who show how the written and spoken words may be more efficiently organized to successfully transmit messages and how the three-dimensional world may be coded on the plane surface of a paper through systematic and optimum graphic representations. In chapters four through six Professor Rathbone sets down the results of many years of experience in teaching students and others how to write well in technical areas. His purpose is to combine a pragmatic exposition of how to write, read, and speak more effectively with correlative examples illustrating the principles set down by Professor Rosenstein in the earlier chapters. Professor Schneerer has undertaken to do much the same thing with graphics. He presents a concise description of sketching and the principles of graphic representation which allow one to quantitatively describe a three-dimensional object on a plane. He shows that visual communication is a form of coded data with tremendous information content and he illustrates how information theory can point the way to more optimum and efficient communication in graphics.

The over-all aim of this book is to provide the student with a new conceptual framework for the methods of communication already familiar to him and to build on this framework with specific teachings which will enable him to read, write, speak, and draw more effectively.

During the past five years Professor Rosenstein has served as the Co-Principal Investigator of the UCLA Department of Engineering Educational Development Program. This program with assistance of a grant from the Ford Foundation has been devoted to an extensive study of engineering and engineering education. While it is impossible to adequately credit all contributions, this author wishes to acknowledge the insights gleaned from the discussions and investigations of the E.D.P. Design Subcommittee. Particular mention should be made of the contribution of Deans L. M. K. Boelter and M. Tribus, and Professors D. Rosenthal, J. M. English, M. Asimow, J. Beggs, A. Powell, F. Shanley, J. Lyman, and J. Powers.

Professor Rathbone acknowledges the encouragement and advice given him by Dr. Kenneth Wadleigh, Dean of Students at M.I.T.

Professor Schneerer expresses his appreciation to the many colleagues at Case who have encouraged and supported his efforts to present graphics as a language for the effective communication of technical information.

J. B. RESWICK

Case Institute of Technology

CONTENTS

PART I COMMUNICATION

1. **The Role of Communication in Engineering Design, 3**

    Introduction · The Design Process · Communication and the Design Process · Communication and Information · Model Theory.

2. **The Mathematical Theory of Communication, 11**

    Information · Channel Capacity Meanings · Industrial Communication Criteria · Human Communication Channels · Macroscopic Analysis.

3. **Communication Systems, 32**

    Man-Man Communication · Man-Machine Communication · Communication Skills · References.

PART II COMMUNICATION PRACTICE

4. **The Reader, 41**

    Importance of Feedback · Identifying the Reader · Analyzing the Reader's Channel Capacity · How to Code the Message for Maximum Effectiveness · How to Code a Message for the Rapid Reader · Summary.

5. **The Writer and the Report, 51**

    Why Write a Report? · Writing as a Communication System · Characteristics of the Mechanical Channel · Problems at the Information Source

—Preparation for Writing · Problems at the Encoder—Writing · Quick Tests for Proper Encoding—Editorial Function · Summary.

6. **Oral Reporting, 63**

   Comparison with Written Communication · Suggestions for Improving Input and Output · Visual Aids · Summary.

7. **Graphics in Engineering Design, 71**

   Introduction · Graphic Communications · Graphic Forms.

8. **How to Sketch, 87**

   Tools of Sketching . Sketching.

9. **Techniques of Sketching**

10. **Charts, Graphs, and Mathematic Constructions, 109**

    Charts · Graphs and Mathematic Constructions.

11. **Presentation, 122**

    Report Illustrations · Slides · Posters · Summary · Bibliography

ENGINEERING

COMMUNICATIONS

part I | COMMUNICATION THEORY

chapter 1 | THE ROLE OF COMMUNICATION IN ENGINEERING DESIGN

INTRODUCTION

This book has been written to illustrate the role of communications in the professional activities of the engineer. Its purpose is twofold: to give the engineer new insight into the nature and role of communications of all types and to show him how to apply the theory of communication in his day-to-day professional activities. Since design is the essence of engineering, the need and the role of communications in design will be first presented. Following this a foundation for communications of all types will be developed through information theory. Next, the essential characteristics of the important communication means existing in our society will be explored. The application of the theory of communications to the all important problem of interpersonal communications will occupy the latter chapters of the text. Here the basic concepts originally developed for mechanical communications systems will be shown to establish the fundamental rules for effective communications among people.

To effectively present the place of communications in design, and consequently in engineering, it is desirable to have a concise definition of both design and communication along with a clear understanding of both the design and communication processes. As design and the design process is the central theme in this series, let us discuss design by first defining what we mean by the term "design." *Design has been defined as an iterative decision making process which is employed to develop means for optimizing the value of resources.*

The theories of communication have been developed to the point where they can be demonstrated to serve as adequate foundations for the description of any form of communication. In each case we find that communication involves a choice of alternatives or, in other words, a sequence of decisions just as does design. It is not surprising then to find that the basic ideas and methods of the discipline of design closely parallel those of communication theory. In addition, of course, communication forms the sinews that hold the design process together.

THE DESIGN PROCESS

Studies of the process invoked by engineers in designing a variety of systems ranging from electronic controls to airplane wings have shown that regardless of the end object the design process is essentially the same. The basic elements of the process have been collected and called the "Anatomy of Design"—the following steps, which are repeated over and over until a satisfactory product is produced.*

Anatomy of Design

1. *Identification of the needs (i.e., defining the problem)*. A careful analysis of the extent and validity of each need is required.
2. *Information collection and organization*. All factors which relate to the system need to be considered. Where necessary, experiments must be devised to obtain data otherwise unavailable.
3. *Identification, modeling, and statement of system variables*. All factors influencing the system—the so-called "boundary conditions."

    Engineering systems (and, in principle, subsystems and components) can be analyzed into basic elements which, when described or prescribed in appropriate detail and when properly synthesized, will constitute the design of the system.

    a. Inputs: Those resources and other environmental factors which are converted (or modified) by the system in question.

    b. Outputs: That which is produced by the system.

    c. Transforming means: The device used to obtain the relationship between inputs and outputs.

    d. Constraints: All elements and factors which express limitations and/or need to be accounted for in the design.

---

* The basic elements of the design process are listed in the "Anatomy of Design." "Design as a basis for an Engineering Curriculum," (Allen B. Rosenstein, J. M. English, *Proceedings of the first conference on Engineering, Design Education,* Cleveland, Sept. 8, 9, 1960, pp. 1-28.) The detailed *sequence* of steps described in order of occurrence has been called the "Morphology of Design." (Morris Asimow: *Introduction to Design;* Englewood Cliffs, N. J.: Prentice-Hall, Inc., 1962). Taken together, the "Anatomy" and the "Morphology" offer a complete description of the design process and all of its elements.

4. *Criteria development for optimum design.* The rules for judging relative merit.
    a. Development of value system.
    b. Criteria relationship among values.
5. *Synthesis.* Evolving of systems to convert the inputs into the desired outputs. At this step only the requirement of realizability is usually met.
    a. Physical realizability.
    b. Economic worthwhileness (realizability).
    c. Financial feasibility (realizability).
    d. Realization of producibility.
    e. Realization of reliability.
    f. Realization of maintainability.
6. *Test evaluation and prediction (analysis) of performance.*
7. *Decision steps.*
8. *Optimization.* (Maximizing the performance. Reduction to "best" system with available knowledge.)
9. *Iteration.* It is recognized that the above operations are found throughout the design process. Many iterations will be taken around several or all of them. In particular, the engineer continually re-examines his previous findings and decisions in the light of new information.
10. *Communication, implementation, and presentation.*

COMMUNICATION AND THE DESIGN PROCESS

The Anatomy of the Design Process can serve as an excellent outline for our study of communications. The basic problem lies in the need of the engineer to communicate with other men, with machines, and with himself. Without highly developed, and in many instances highly sophisticated, communication channels and techniques, the engineer is rendered helpless. If his incoming communications are blocked, he becomes sterile. The problems available to him become restricted and his knowledge of new resources becomes limited. If his outgoing communications are reduced, he is shortly reduced to futile frustration. For again, an engineer who cannot effectively communicate his design results or who cannot quest for new knowledge will be of little practical value. Let us discuss the major aspects of engineering communications in terms of some of the elements of the anatomy of design.

Definition of the Problem. This is normally the first step in any design.

The engineer may receive the problem in the form of a set of specifications that must be satisfied or he may be confronted with the physical situation itself. In any event, a major concern is the translation of some situation in the real world into a suitable model or representation that can be studied and manipulated. This translation can be effected by means of words, pictures, sketches, graphs, equations, audio-visual means such as motion pictures, and/or even scale models.

Information Collection and Organization. The merchandise or medium of exchange in communication is knowledge or information. We shall use the word information and, in developing Information Theory, shall give it a quantitative definition. Consideration of information leads us to the second element in the Anatomy of Design. Since design is a decision-making process and decision denotes a choice between alternatives, a design project of any magnitude entails the gathering and organizing of substantial amounts of information. In addition to the information-transmission problem that is beginning to take shape, the reader should now begin to comprehend the need for systems of information storage and techniques for information organization and retrieval.

Of course, substantial blocks of technical information are stored in men's minds. In addition to these living information storage bins, engineers have found that information can be stored as well as transmitted by pictures, sketches, written reports, etc. In recent years engineers have greatly extended man's ability to collect, store, and organize information through the use of computers and their magnetic memory drums, tapes, and cores.

Development of a Value System. Unique answers to substantial engineering problems are obtained only in light of the value scales of contemporary society. By changing the value system, we can change the correct answer to almost any engineering problem. The task of developing communications and applying the value system often requires more effort than the pure technical aspects of a problem.

Synthesis and Optimization. We consider the *synthesis* of a *possible* answer complete when the proposed solution is shown to be physically, financially, and economically realizable as well as being producible, reliable, and maintainable. The *optimization* step is complete when the "best" solution in light of our value system is produced from the array of realizable solutions.

Communication, Implementation, and Presentation. The study of the communication, implementation, and presentation of a design is the concern of this text. Again let us state that a design that cannot be implemented, cannot be tested, or cannot be communicated might just as well not exist at all. Unfortunately, it is just in these very stages that many excellent pieces of engineering have been defeated or depreciated.

COMMUNICATION AND INFORMATION

Any corporation, business, or for that matter any human organization, is in reality a communication system. Its effectiveness at all times is dependent upon the efficiency with which it communicates. To more fully appreciate this we shall first present an operational definition of the word "communications." We will then develop a quantitative understanding of the word "information" and a better understanding of the elements of a communication system.

Communication

The theory which is to be developed is dependent upon the initial assumption and understanding of a very broad operational definition of the word "communications." For purely mechanical systems, communication has been called the *"process by which one mechanism affects another."* S. S. Stevens has proposed as an operational definition that, *"Communication is the discriminatory response of an organism (object or mechanism) to a stimulus."* [1]*

These definitions tell us that a communication is initiated only to effect some predetermined result and has failed if the desired specific result is not achieved. This means that the message that gets no response, or the wrong response, is not a communication. If communication is evaluated on the simple basis of intent and result with a yes-or-no (success-or-failure) scale, quantitative results can be obtained for a diverse span of information transmitting systems.

We are accustomed to think of communication in terms of human-to-human communication with perhaps some mechanical transmission links. In the industrial system, a high percentage of the communications are human-to-machine. When a production scheduler writes a work order for punch press #52 to produce 100 individual stampings on a given day, he is essentially communicating with the machine. The departmental foreman and the machine operator are simply links in the chain between scheduler and machine. If the machine produces 98 parts instead of the desired 100 individual stampings, the communication received by the machine and transmitted to the raw material was 2% in error. The original message called for 100 successful machine operations. Only 98 operations got through. It is considered that of the original 100 part message, two parts were in some manner lost in transmission. Once the concept of human-to-machine communication is accepted, machine-to-machine com-

---

* Superior numbers pertain to references grouped at the end of this part.

munication appears logical and it then follows that in some portions of our industrial communication system, machines will be directing humans.

Information

The terms "information theory" and "communication theory" are frequently associated and often implicitly implied to be synonymous. Modern information theory as developed by Claude Shannon was the outgrowth of mechanical communication system studies. To develop quantitative theories for information handling communication channels, it was first necessary to define and measure *information*. This was accomplished by a statistical definition. Information is defined statistically in terms of not so much what you say but what you could say. Information is a measure of one's freedom of choice when one selects a message. If there were only one message that could be sent, the information would be zero as there would be no choice and no point in sending the message. With information quantitatively defined, the characteristics of mechanical communication channels can be developed along with other factors, such as noise, which influence information flow.

From its early mechanical system applications[2,3,4,5] the use of information-communication theory has grown until we now find its principles being applied to such diverse fields as psychology,[6] psychiatry,[7] biology,[8] economics,[9] and linguistics.[10,11] It is particularly interesting to note that areas are included that in the past have presented major obstacles even to qualitative analysis.

The power of information-communication theory can be properly evaluated by considering the nature and scope of the mathematical problem that is presented and solved for the mechanical communication system. The theory is specifically designed for non-linear systems subject to random perturbations (noise) with attendant error probabilities. These systems particularly have limited transmission capacity and exhibit very sharp saturation characteristics, i.e., all information transmitted at rates above the system capacity will be in error. The theory is further designed to analyze information transmitted in continuous messages such as the human voice or by discrete symbols such as teletype or Morse code. In the industrial organization, we are normally concerned with the movement of discrete objects through a manufacturing system with a finite capacity. Both the information and the objects moving through the industrial system are exposed to varying amounts of noise and consequently are always subject to some finite error probability.

MODEL THEORY

S. S. Stevens has said, "The stature of a science is commonly measured by the degree to which it makes use of mathematics. Yet mathematics is not itself a science, in the empirical sense, but a formal, logical symbolic system—a game of signs and rules. The virtue that makes mathematics more than trivial is its capacity to serve as a model for events and relations in the empirical world. Like any model used to represent something other than itself, mathematics 'fits' better in some places than in others, but at no place is there perfect correspondence between the mathematical model and the empirical variable of the material universe. Generally speaking, the fit is better to the degree that the dimensions and qualities of the things we study are measurable on well-founded scales." [12]

Mathematical and/or conceptual models of physical systems are commonplace tools of the engineering analyst. Intricate mechanical, biological, or social organizations often can only be mastered and properly understood when a final conceptual model is constructed with a reasonable mathematical "fit." The evolution of this model normally progresses from the original qualitative system analysis to the determination of the mathematical form of the system processes, and finally results in the selection of an initial mathematical model and set of operational hypotheses. Since it is often too difficult to state directly whether or not the postulates correspond with reality, a testing program is necessary to check the computed performance of the model against the known responses of similar real organizations. In this manner a measure can be had of the validity of the original assumed hypotheses. Where the "fit" of the model and the actuality is intolerable, the model postulates are repeatedly adjusted and retested until the model predicts with the desired degree of accuracy the general real system performance. The words "desired degree of accuracy" obviously require individual situation interpretation, since for some systems a model giving only a good qualitative understanding of a process would be a major achievement.

The mathematical "fit" of communication theory to industrial organization and engineering is good. With the acceptance of the industrial organization as a communications system, the performance of the mechanical components of this humanistic-mechanistic system can be fully described by communication theory. The reader realizes, of course, that the problems of engineering design are embedded within the industrial organization. The problems of engineering communication are essentially the same as those found throughout the industrial system.

The human portion of the humanistic-mechanistic system presents practical, but not theoretical problems for the application of the theory.

A human machine's ultimate transfer characteristics should be theoretically expressible in terms of his purely mechanical characteristics. Suppose in a mechanistic industrial communication model the human components are replaced by equivalent mechanical components with identical physiological characteristics. The model would then predict the best possible performance achievable from that industrial system. Since the machine replacements of the humans would possess all the human capabilities including memory, but excluding all emotions, the model would predict a performance unimpaired by the noise introduced into the real communication system by the emotional conflicts of humans. Human emotions thus add to the real industrial organization an additional degree of unpredictability, with a resultant predictable probability of increased system error.

In turning to the psychologist for aid in fitting the psychological characteristics of the human elements into our industrial communication model, we find that the psychologist has already discovered and applied information theory. David Grant writes, "In spite of limitations there will be extended a profitable use of information theory in the future of psychology. Stimuli are statistical in nature; so are responses; they are related—they form a communication system single or multichanneled, with or without storage capacity. Psychology is the study of this system. Shannon's (information) theory is a most valuable tool in this study."[6]

Before considering any engineering application let us first examine the definitions and theorems of the mathematical theory of communications.

chapter 2

THE MATHEMATICAL THEORY OF COMMUNICATION*

By presenting now the results and implications of the mathematical theory of communication without undertaking detailed theorem proofs, the engineer's questions of "Why is it?," "What is it?," and most important, "What is in it for the engineer?," can be answered.

The "why" of a theory of communication lies in the fact that the key problem of communication is one of statistical inference. In any system, the message received is never a mathematically exact replica of the original transmission. In transmission, the message will be subjected to random unpredictable perturbations commonly called "noise." The recipient of the message must apply statistical inference in attempting to extract the original message. Since all communication systems have some error potential, they are best described in the mathematics of statistical inference. As the "what" of communication theory is presented, the statistical nature of each element will be discussed.

The basic elements of a mechanical communication system are given in Figure 1. The information source is the device, person, persons, etc., which

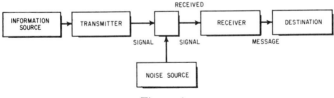

Figure 1.

in order to produce a desired change in the state of the destination, selects a desired message from a set of possible messages. The transmitter changes the message into a signal form which can be conveniently trans-

---

* As developed largely by Claude Shannon for mechanical communication systems and with concepts from Shannon and Weaver,[3] Sanford Goldman,[2] and Brockway McMillan.[6]

mitted by the channel. We say that the transmitter encodes the message for transmission. The channel is the machine over which the signals representing the message are transmitted to the receiver. The receiver decodes the signals into the message normally in a form meaningful to the destination. Noise is said to be any additions to—or distortions of—the original message. It is obvious that noise may take many forms. However, noise is essentially any *deviation from the original source intent and consequently noise might well be called "error."*

The block diagram of the mechanical communication system can be readily illustrated by a great number of physical situations. Since Shannon's work was sponsored by the Bell System, the names given the elements of the block diagram are fully descriptive of the situation of telephonic communication between two individuals. In this case, at least, the concept of noise is easily related to one's personal experience. Again note that the communication engineer considers any change from the original message as noise. The attempts to answer the problems presented by this random statistical phenomenon of noise or error have produced our modern communication theory. With the recognition of noise, the extent of the communication task is further clarified. The receiver or decoding device must not only decode the noisy signal received, but it must also infer what the original signal really was.

Along with the noise that nature and man introduce into his communication channel, the engineer must concern himself with the additional all-important statistical problem of bulk or volume of channel message traffic. The message bulk will be intimately related to the encoding process. One can easily verify experimentally that a long sample of English text can be readily restored to its original form even if half its letters or all of its vowels are deleted. Since in English all letters do not occur with equal probability, e.g., the letter u always follows the letter q, and because of the unique meaning inherent in a long passage, there exist strong statistical connections extending over stretches of message. When a channel transmits English messages, its true information content is reduced because part of the message is devoted to telling us something we already know—the probability structure of the English language. All that is really needed is enough key data to allow a unique restoration of the original message. All other material would be redundant and a waste of channel capacity.

The latter part of the above sentence is not entirely true, however, for redundancy and noise interact. In a noisy channel where some of the message will probably be received in error, redundancy can be effectively used to reduce the final message error to any desired fraction. A common example of the manner in which the final decoder uses redundancy can be seen in the ease with which we normally perceive the correct message even in badly misspelled telegrams.

Shannon has posed and answered four questions about the general communication problem:[3]

1. How can the amount of information and rate of information production be measured in a manner which will take into account the probability structure of the message?
2. How can the communication channel "capacity" be measured? How many information units per second can be sent through a channel with a given signal power and a specified noise?
3. What are the characteristics of an efficient encoding process? With maximum encoding efficiency, at what rate can a channel convey information?
4. What are the general characteristics of noise and their effect on final message accuracy? How are the undesirable effects of noise minimized?

INFORMATION

In the following section, while we are concerned only with the mechanical aspects of communication, information is not to be confused with meaning.* One's initial contact with information in communication theory usually seems strange, for here information relates not to what is said, but to what could be said. Information is a measure of one's freedom of choice when one selects a message. As we said earlier, if there were only one message that could be sent, the information would be zero as there would be no choice and no point in sending the message.

Because information is going to be measured in terms of the number of

---

*Shannon's Information Theory concerns itself with the problem of the transmission and detection of information. The detection problem is essentially one of determination of the original information or message. L. Brillouin (*Science and Information Theory*, New York: Academic Press, Inc., 1956.) points out that Information Theory ignores the value of the information handled by the system. This element of information value, however, is needed whenever one considers information as a basis for prediction or decision. It is the Theory of Statistical Decisions which is concerned with the meaning and use of information. We see, therefore, two related but different decisions. The first concerns the determination of the information that has been sent through a communication channel and here Information Theory helps us. The second involves a determination of the proper course of action in light of given information. Here the Theory of Statistical Decisions is of some help.

possible choices, it has been found that quantity of information may be conveniently expressed in binary code or, in other words, measured in terms of $\log_2$. For example, if one had 16 alternate messages from which he is equally free to choose, then $16 = 2^4$ and $\log_2 16 = 4$. This means that the situation and all its particular messages could be characterized by 4 "bits" of information.

This idea of the *quantity of information* can be illustrated by a simple parlor game. Suppose we write a number on a piece of paper and ask you to determine the number with a *minimum* number of guesses. We tell you, for example, that the number is somewhere between one and thirty-two and that we will give you a binary answer, i.e., *yes* or *no*, to each of your questions.

Information theory tells us that any situation involving 32 equally probable choices can be characterized by $\log_2 32 = 5$ bits. Therefore, the answer should be obtained with a maximum of five questions. For example, if the number were 25, the game might go:

1. Is the number greater than 16? (i.e., $\frac{32}{2}$)—answer: yes.
2. Is the number included in 17 to 24? (i.e., $16 + \frac{8}{2}$)—answer: no. (Therefore, it is within 25 to 32.)
3. Is the number included in 25 to 28?—answer: yes.
4. Is the number included in 27 to 28?—answer: no. (Therefore, it is 25 or 26.)
5. Is the number 26?—answer: no.
(Therefore, it is 25.)

By always dividing the remaining choices into two equally probable groups, you would have determined the answer to be 25 in no more than five guesses. We say, then, that it takes a minimum of 5 binary bits of information to describe any message that might be chosen from 32 equally probable messages.

We are really using the idea of information when we speak of an engineering organization or an industrial organization. The purpose of *organization* is to maximize productivity by reducing the number of choices involved in each decision. A group is said to be highly organized if each man knows exactly what to do in order to resolve the questions which confront him. If a situation is highly organized, it is *not* characterized by a large degree of *randomness* or *choice*—that is to say, that the information is low.

Maximum choice or *information* for a given set of symbols will occur when the probabilities of occurrence of each symbol are equal.

"*Relative information*" for a given situation is defined as the ratio of the actual information (freedom of choice) to the *maximum possible information* for the same set of symbols. "*Redundancy*" is defined:

$$\text{Redundancy} = 1 \text{ minus relative information} \qquad (1)$$

"Redundancy" then is the fraction of the message that is determined not by the choice of the sender, but rather by the statistical rules governing the use of the symbols in question. As was pointed out, this fraction of the message could be omitted in a noiseless situation and the message would still retain full information. English has been proven to be approximately 50% redundant.

Information H is mathematically expressed for the case of interdependent symbols whose probabilities of choice are $P_1 \cdot P_2 \cdots P_n$ by the equation:

$$H = -\sum_i P_i \log_2 P_i \qquad (2)*$$

The minus sign only appears since P_i is usually less than one and H is taken as a positive quantity. If the probability of any symbol is unity (symbol is certain to appear each time) $\log 1 = 0$ and the information is zero since only one symbol will always be transmitted. The units of H are bits per symbol.

"*Channel Capacity*" is defined in terms of the amount of information correctly transmitted per unit of time. Bits per second is the common unit. For example, a teletype channel could have a source that may freely choose from among 2^s symbols. Each symbol chosen then represents $\log_2 2^s = S$ bits of information. If the channel is mechanically capable of correctly transmitting n symbols per second, then the capacity C of the channel would by definition be ns bits per second.

At this time, we have defined information H in terms of bits per symbol. Information can also be rated in bits per second, in which case the symbol H' will be used. Without stating how we would determine the capacity C of a real channel, C has been defined in bits per second. With this rather elementary background the fundamental theorem of communication can be stated in all of its simplicity.

Fundamental Theorem of Communication

"By proper coding procedures it is always possible to transmit symbols over a channel of capacity C, when fed from an information source of H bits per symbol at an average rate of nearly C/H symbols per second with zero error. However, no matter how clever the coding, the errorless transmission cannot exceed C/H."

---

*This form is similar to the defining equation for entropy in statistical mechanics. The term entropy has been adopted in classical information theory for the measure of uncertainty or information. Thus information theory is linked to the thermodynamics of Boltzman and Gibbs with the result that the mathematical methods of statistical mechanics form a guide for information theory development.

The fundamental theorem can be stated another way. If the information source generates at a rate of H' bits per second and feeds a channel of capacity C then errorless transmission can be approached if $C > H'$.

Information has been defined in terms of the freedom of choice in selecting a message. The greater the freedom, the greater the information, and the greater the uncertainty that the next message to be received will be some particular one. "Noise" causes the received signal to exhibit greater uncertainty or more apparent information than the transmitted signal. Noise, therefore, is undesirable uncertainty or information which tends to confuse the decoder and cause errors. The powerful fundamental theorem states rather simply, however, that no matter how much noise is present, a coding method will exist that shall allow errorless transmission at a rate H' which shall approach C, the channel capacity. The statement of errorless transmission in the face of error producing noise is provocative. The only possible joker in the erstwhile simple statement might lie in the still unexplained derivation of C. (It so happens that C will be dependent upon the amount of noise present in the channel as will be subsequently shown.)

Illustrative Example

By illustrating the fundamental theorem with an example, it is possible to tie together the implications of noise, channel capacity, rate of information transmission, redundancy, transmission accuracy, the coding problem, and coding time. Consider the following elementary situation: A channel is capable of mechanically transmitting 100 numbers per second. However, the noise in the channel is great enough to insure that one (and only one) number out of each 100 transmitted is received in error. If this is an intolerable accuracy, we immediately look for some means of improving the transmission accuracy through this noisy channel, assuming we have no control over the channel and its error producing rate. The first and most obvious approach would be to catch the errors by introducing message redundancy into the communication. If each number transmitted were repeated before the next number were transmitted, any number transmitted in error would immediately show up as it would not match its preceding (or following) mate. Thus, by introducing 50 per cent redundancy into one's message, we would immediately spot any error when the second number of the pair were received. This simple error-detecting redundancy has the advantage of speed. It has the disadvantage of 50 per cent reduction in original transmitting capacity, plus the inability to tell which of the unmatched pair is correct.

By transmitting each number three times, 66.6 per cent redundancy, the transmission can be made 100 per cent accurate. However, the useful capacity of the system would now be only 33 numbers per second. The fundamental theorem, though, would lead us to expect much greater capacities for the system.

Let us try a slightly more sophisticated coding scheme with controlled redundancy. Suppose the numbers were transmitted in the sequence indicated in the following array:

$$\begin{array}{cccc} a_{11} & a_{12} & a_{13} & T_{R_1} \\ a_{21} & a_{22} & a_{23} & T_{R_2} \\ a_{31} & a_{32} & a_{33} & T_{R_3} \\ T_{C_1} & T_{C_2} & T_{C_3} & \end{array} \qquad \begin{array}{l} T_{R_1} = a_{11} + a_{12} + a_{13} \\ \\ \\ T_{C_1} = a_{11} + a_{21} + a_{31} \end{array}$$

Here the a's are the original transmission numbers and the T's are the controlled redundancy with T_R's equalling the sums of the rows and T_C's equalling the sums of the individual columns. For every 15 numbers transmitted, 6 are redundant for a redundancy of 40%. The code is self-checking. An error in any one single number in the array can be found and corrected by the time T_{C_3} is received. Now if a 9×9 array were used instead of a 3×3, 81 useful numbers would be transmitted along with 18 redundant check figures. In this case the redundancy is only $\frac{18}{99} = 18\%$.

The useful error-free transmission capacity has been increased to 81.0 numbers per second, but it is now necessary to wait for the transmission of up to 99 symbols in order to read the complete message, i.e., find and correct the error that must be present in every 100 numbers transmitted.

Channel Capacity

The heuristic proof embodied in our example bears out the fundamental theorem. By clever encoding we are able to use controlled redundancy to cause the error-free transmission rate of the system to approach the capacity of the channel. However, this increase in rate is not obtained without sacrifice, for the waiting time for message decoding (correction) increases rapidly as the transmission rate approaches its theoretical error-free maximum.

The ultimate error-free capacity of a noisy channel has been developed by Shannon for the transmission of both discrete and continuous symbols.

Capacity C of a discrete channel:

$$C = \text{Max } [H'(y) - H'_z(y)] \qquad (3)$$

where $H'(y)$ is the total information including noise in received signal,

$H'_z(y)$ is the information (uncertainty) contributed to received signal by noise.

The maximum is with respect to all possible information sources used as input to the channel. For a noiseless channel, $H'_x(y) = 0$. Note again that the fundamental theorem says that for $H' > C$, the minimum error regardless of coding cannot be less than $H' - C$. Equation (3) lends itself to an experimental determination of C. Over an extended period of time, a record can be kept of the messages sent and the results achieved (messages received). At a later time a symbol comparison can be made and the probability of each received signal re-evaluated and averaged. Where the received symbol coincides with the message symbol, the probability is one and information is zero. $H_x(y)$, the uncertainty of the received signals, will be left. Shannon and Weaver offer further examples to illustrate the idea of capacity of a discrete noisy channel.[3]

The engineer may apply the discrete channel theory to a variety of problems ranging from the production line to teletype systems where he usually deals with discrete objects or symbols. However, there will be many occasions where the continuous messages such as the speaking voice will be used. The continuous signal, maximum capacity theorem is especially important for the light it throws on the relations between capacity, signal strength, and noise magnitude.

Continuous signal maximum capacity theorem:

$$C = W \log_2 \frac{P + N}{N} \text{ bits/sec} \qquad (4)$$

where C is the maximum capacity of channel, in cycles per second
    W is the frequency band width
    P is the average signal power
    N is the average noise power.

The theorem indicates that with zero noise, the channel capacity becomes infinite. With efficient coding, information could then be transmitted at an infinite rate. Unfortunately, there are no truly quiet places on our earth. In every physically realizable system some noise will exist and C will be finite. The formula further indicates the well-known threshold concept, for as N approaches and exceeds P in magnitude,

$$\frac{P + N}{N} \to 1 \quad \text{and} \quad \log_2 \frac{P + N}{N} \to 0$$

or zero useful capacity.

At this point we have completed our simplified presentation of the mathematics of communication as developed for mechanical systems. Without further expansion these mathematics can be used to describe and predict the ultimate performance of a humanistic-mechanistic industrial organization with the people replaced with perfect "man machines." To account for the somewhat less ideal performance of the actual system, it is not necessary to modify the mathematics, but only to recognize the additional error probabilities introduced by the human elements. This can be best accomplished by rapidly previewing the all-important concept of channel capacity.

CHANNEL CAPACITY MEANINGS

A channel for transmitting code messages is a complete transmission system from an input location to an output location. The properties of the channel include the properties of the equipment used in the channel as well as the properties of the codes and languages used. Channel capacity can usually be specified at least three ways; in terms of code capacity, language capacity, and semantic capacity. The code capacity of a channel is determined by the fixed constraints which may be part of the code channel (for example, the dot-dash-space of electrical telegraphy) plus all the fixed constraints which may be present in the letter, word, and sentence codes of the language employed in the channel. The language capacity of a channel is obtained by superimposing the probability constraints of the language upon the fixed constraints of the code channel. A language such as English thus represents a second coding that we impose on top of the mechanical code of the channel.

It can be proven that the language capacity of a channel will be less than its code capacity unless the probability of occurrence of a symbol is properly matched to the time duration of its code transmission. For example, the letter *e* being the most frequently employed in English would require the shortest code representation.

Language capacity as used herein is defined in terms of symbol by symbol (i.e., letter) transmission. In the mathematics of the mechanical transmission system, no attention is given to the meaning of the language symbols since meaning has no influence in the design of the system.

When we become concerned with the *meaning* of the message transmitted, we then consider semantics. Semantic information and, consequently, semantic capacity can be defined by the basic information definition:

$$\text{Information received} = \log_2 \left[\frac{\text{Probability at the receiver of the event after the message is received}}{\text{Probability at the receiver of the event before the message is received}} \right]$$

The numerator accounts for the noise in the system, for in the noiseless case:

$$\text{Information received} = -\log_2 \left[\text{Probability at the receiver of the event before the message is received} \right]$$

We might illustrate the above with the following example: Suppose 16 sailboats of presumably equal design and equally capable crews were to race from California to Hawaii. The judges in Hawaii will send us a simple telegram giving only the number of the winning boat. The boats are numbered from 1 through 16. Since the boats have equal probabilities,* the *probability at the receiver of the event before the message is received* $= \frac{1}{16}$. Consequently, in the noiseless case the message gives us:

$$\text{Information received} = -\log_2 \frac{1}{16} = 4 \text{ bits of information}$$

Now, however, suppose we were aware of the fact that the telegraphic system to Hawaii was not 100% dependable. We all have had experience with situations where numbers had been jumbled in telegrams. We, therefore, would still have some reservations about the true winner of the race. If we knew the reliability of the telegraphic system, we could compute the information received. If, in this case, we knew the system produced an error in numbers 1 to 16 about once in every 32 messages, we would have to compute the information received for the two possible cases, i.e., (1) the message is correct, or (2) the message is incorrect.

Case (1): The message is correct.

Since any boat can win:

$$P_A = \text{probability before reception} = \frac{1}{16}$$

But, since one out of 32 messages is incorrect:

$$P_B = \text{probability after reception} = \frac{31}{32}$$

$$\text{Information}_{\text{correct message}} = +\log_2 \frac{\frac{31}{32}}{\frac{1}{16}} = \log_2 15.5 = 3.95 \text{ bits/message}$$

Case (2): The message is incorrect.

Suppose number 14 is sent and number 8 is received.

$$P_A = \text{probability before reception} = \frac{1}{16}$$

After the message is received we have to compute the probability of number 14 being sent if number 8 is received. First, the probability of error is $\frac{1}{32}$. Then, if the transmission is in error, the remaining numbers have equal probability of being correct, or $\frac{1}{15}$.

---

*The equation does not depend upon situations having only equally probable alternatives. To apply the equation we need only an estimate of the probabilities of occurrence of each individual event.

$$P_B = \text{probability after reception} = \frac{1}{32} \times \frac{1}{15}$$

$$\text{Information}_{\text{incorrect message}} = \log_2 \frac{\frac{1}{32} \times \frac{1}{15}}{\frac{1}{16}} = \log_2 \frac{1}{30}$$

$$= -\log_2 30 = -4.9 \text{ bits/message}$$

The *average* information per message is dependent on the relative occurrence of accurate and inaccurate messages.

$$\text{Average information/message} = \frac{31}{32} \log_2 15.5 - \frac{1}{32} \log_2 30$$

$$= 3.68 \text{ bits/message}$$

Goldman[2] has discussed language and semantic capacity in some detail. Weaver[3] points out that the introduction of the semantic problem causes a modification of the rather simple diagram of Figure 1 to recognize that a "semantic receiver" is interspersed between the mechanical receiver and the destination (Figure 2). This semantic receiver performs a second

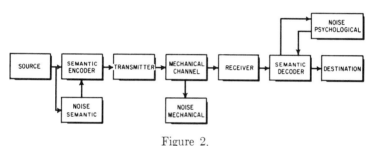

Figure 2.

decoding upon the message for it must match the statistical semantic characteristics of the message to statistical semantic capacities of the receivers. A semantic encoder exists between the information source and the transmitter to encode the intended meaning of the sender into symbols that the receiver can satisfactorily decode. In addition, a box must be introduced at the source entitled semantic noise to account for perturbations or distortions of the meaning not intended by the source, but which affect the destination.

The semantic problem of the engineering situation is enormous. Undoubtedly a substantial portion of the difference between ideal and actual engineering performance can be traced to semantics. Adequate semantic channel capacity requires not only a common language, but a good vocabulary match, and an extensive sharing of cultures and experience. Here the meaning of the word semantics must be expanded to include not only the meaning of words, but the meaning of situations, and experiences. The term semantics is used here to imply meaning, not just the dictionary meaning of the words transmitted, but meaning in terms of the original "intent" of the sender.

The dependence of channel capacity upon this expanded view of semantics is illustrated by the oft paraphrased story of a college professor who, lost in the mountains, asked a hillbilly for directions.[13] After the professor indicated he didn't know where Knudsen's bridge, Gin Creek, or Ma Perkin's hog wallow were, the mountaineer gave up in disgust, saying, "You can't tell nuthin' to nobody who don't know nuthin' to begin with!"

He was correct, for in terms of the mountaineer's message, the professor's semantic channel capacity was zero. They both spoke and understood the same language, but lacked the common object identification experiences which would have allowed them to communicate. Although the recognition of semantic mismatch between source and channel is frequently difficult, once the problem is recognized, it is relatively simple to evaluate in terms of communication theory.

Psychological Noise

The introduction of the effects of the emotional characteristics of the human components represents the last adjustment that must be made to the mechanical communication system to faithfully simulate the engineering organization. Human emotional factors normally manifest themselves in the form of extraneous noise introduced between the semantic decoder and the destination (Figure 2). Because of its psychological origin, we have chosen to call this noise "psychological noise." "Psychological noise" is intimately associated with what other writers have called the "effectiveness" problem.[6]

Psychological noise can be activated either by the message itself which may disturb the receiver or by events outside the work place. Emotional factors can also cause additional semantic noise. A subordinate will frequently attempt to encode a message so as to both convey information and still please his superior—a situation usually not conducive to accurate communication. The effects of psychological noise are the same as those of mechanical noise. The useful channel capacity will decrease as the noise increases. The student might evaluate the effects of psychological noise by deciding how much information he would extract from his studies if he had *just* learned that his fiancee had eloped with his best friend. While the system performance with varying amounts of psychological noise is readily understandable qualitatively, much experimental work must yet be performed before quantitative results are available. The extreme change in performance that can be observed in a normally efficient worker under severe emotional stress is not surprising when considering the channel capacity expression:

$$C = W \log \frac{P+N}{N} \tag{5}$$

where P is the power of the message signal
N is the magnitude of the noise.

With the introduction of semantic and psychological noise, the application of the mathematical theory of communication and its concepts to the industrial communication chain can now be studied in some detail.

INDUSTRIAL COMMUNICATION CRITERIA

Stated simply, it is the responsibility of industrial management to develop the most efficient, rapid, and accurate system possible for the transmission of the customer's request (order) from the sales department through the plant to the shipping department. In other words, the industrial organization is simply a complicated communication system which enables the customer to communicate with the shipping clerk. The same criteria can be restated for any organization. The organization exists to convert the input command into the desired output.

Information theory can be applied industrially on both a microscopic and macroscopic basis. Microscopically, communication theory data and techniques are available to give numerical answers to a potentially wide variety of problems involving the individual human machine. Macroscopically, communication theory enables the engineer to assume the proper perspective in evaluating the overall communication system. He is able to translate the requirements of good humanistic-mechanistic communications into engineering terms and properly evaluate qualitatively, if not quantitatively, the relative effects of each factor. Knowing all the requirements for accurate communication, he can in complicated systems experimentally determine the system parameters. At the simplest evaluation level he can avoid those situations unfortunately so common in industry where accurate communication is literally impossible.

On either a microscopic or macroscopic level, the method of system design is deceptively simple. First the maximum rate, H', at which information must be transmitted is computed usually in units of bits per second. Then the maximum rate at which the proposed system can transmit information is determined using

$$C = W \log_2 \frac{P+N}{N}$$

for continuous waves or

$$C = H'(y) - H_x(y)$$

for discrete symbols. If the capacity is greater than the required transmission rate, then the designer is in luck. He has only to decide on the amount of controlled message redundancy he will allow for maximum clarity. If $H' < C$, then the designer is in trouble. He has the choice of obtaining a better channel by decreasing noise, increasing signal, finding better facilities, or of giving up entirely.

HUMAN COMMUNICATION CHANNELS

On a microscopic level the analysis of the information capacity of the mechanical machine has been so completely formalized that we shall consider only the problem of the human machine. Even in the most menial tasks, the human machine is employed only in situations where the operations require some decision making. In ordinary time and motion analysis, the industrial engineer wishes to minimize the time required for a combination of physical movement time and decision time. Decision time is the lag between stimulus and response.

Time and motion analysis may be considered a study in human information generation and transmission. Man's information transmission capacity is limited by the fixed constraints of the human motor system and by the kind of coding required for individual tasks. As in all information handling systems, transmission efficiency is a function of coding. Maximum human information handling capacity requires the selection of an optimum stimulus code, the selection of an optimum response code, and finally the matching of the two coding schemes.

The capacities of various human information handling channels should be of fundamental interest in time and motion analysis. The average typist can type 60 words per minute without error. With each word averaging five letters, this comes to 300 letters per minute. If the typist were producing nonsense letters, the information contained in each letter would be $\log_2 26 = 4.7$ bits/letter. When the constraints of the English language are applied extending to groups of eight letters with English occurrence frequencies and interletter constraints, the information per letter drops to 2.35 bits. Thus, the average typist produces $(2.35 \times 300)/60 = 11.75$ bits per second while an expert at 120 words per minute can convey 23.5 bits of information per second. A superior telegrapher can also produce 60 words per minute. The stenotype operator using a more efficient coding system with less redundancy can readily transmit 120 words per minute or 23.5 bits per second. Court reporters can double this and handle 47 bits per second. These preceding figures were all based upon random words. In English text where there are substantial interword constraints, the information per letter is approximately one bit per letter or 5 bits per word.

The linguist describes speech with a little more than 30 elements called phonemes. If each element were independent and equally probable, they would each produce five bits of information. A speaker producing random phonemes at the rate of 10 phonemes per second would generate 50 bits of information per second. With interword constraints this would amount to about 25 bits per second.

The above human channel capacities appear very low when compared to those of, say, an ordinary noisy radio channel with bandwidth of 5000 cycles per second, a signal to noise power ratio of 1000, and a resultant capacity of

$$C = 5000 \times \log_2 1001 = 50,000 \text{ bits/second}$$

However, even in the seemingly relatively simple human information channel represented by the typist, information is visually received, decoded in the typist's mind, and then recoded into impulses to be transmitted by the typist's fingers to the paper.

Information handling capacity in the human machine will vary tremendously depending upon the coding match and the human channels utilized. Where maximum response rates are desired, the conduction delay of the nerve channel selected can be important. The impulse conduction velocity of nerve fibers ranges from 100 meters per second down to less than two meters per second. These conduction speeds make the average reaction time of 200 milliseconds for simple motor tasks seem quite reasonable. Thus for motor tasks restricted to a yes-no response, our human channel can produce only about 5 bits per second. On the other hand, a 500 line television screen would contain 250,000 dots. If the dots have equal probability of being black or white, and if the eye could distinguish all and each dot at a rate of, say, 5 times per second, then the visual information received could be 1,250,000 bits per second. The visual tract is by far the highest capacity human channel. The average reader can read 250 words of text per minute or 21 bits per second. A trained reader can reach 600 to 900 words per minute or an apparent ceiling of 75 bits per second.

Some care must be taken in relying upon the higher figures given above. In many cases communication rates are established without taking into account the reduction in information per word that comes with an extended text. The information handling capacity of the human communication channel has been tested by numerous tasks and found to give consistent results.[14] The fastest human information rates were attained in reading long lists of random words.[15] These rates were computed to be 42 bits per second. Using entirely different techniques, G. C. Sziklai found that visual perception was limited to an information rate of 40 bits per

25

second.[16] On tasks that called for perception rates of 60 bits per second, the operators consistently failed. Quastler[17] points out that people cannot respond more than nine times per second or not over five responses per second when discrimination is required. A second limit exists because people cannot distinguish rapidly among more than 32 equi-probable alternatives without becoming confused. For arm-finger movements such as those of a skilled typist or pianist, Quastler concludes that 25 bits per second is the practical upper limit.

Paul Fitts has described a series of experiments[18] in which the information capacity of the human eye-mind-arm motor system was found to average 10 to 12 bits per second. The information rate of the eye-mind-arm channel for four different tasks, in each of which the physical parameters were widely varied, was found to be very consistent. The results can only lead one to believe that Information Theory will be used as a tool for Time and Motion synthesis. In fact, we understand that Information Theory has been used to design finger-operated keyboards such as that of a typewriter.

As an interesting sidelight on Fitts' work, the author established that the human brain is organized to perform multiple, disassociated tasks without loss of production if at least one task is completely repetitive and if the motor channels involved for the tasks are separate. In checking the result of Fitts' repetitive taping experiment which involved the mind-eye-arm, the operators were found to be able to perform arithmetic problems involving the ear-mind-mouth channel with no loss of taping information.

MACROSCOPIC ANALYSIS

With the application of the basic theorems of information theory, the engineer can reduce the seemingly overwhelming complex intricacies of modern information flow to a reasonably simple mechanistic model, subject to qualitative analysis and the application of round number judgment values.

Recognition that the industrial man is a decision making machine capable of transmitting information accurately at a fixed maximum rate makes the requirements for good organization seem almost ridiculously simple. If a decision is to be made from a large number of real or pseudo-real choices, then that decision represents a large amount of information. Since the individual can only transmit information accurately at a fixed maximum rate, the greater the information content of each decision (or movement) the slower the rate at which decisions can be made. This points out immediately the sweeping necessity for reducing the number of choices available per decision.

We see now that production varies inversely with the logarithm of the number of choices. In the engineering department this concept highlights the need for part, method, and design procedure standardization. On the assembly line, the motion production rate of the workman will increase as the number of different physical movements for a task is decreased. To the personnel director, the concept further emphasizes the need for experience and training. In the average industrial situation, the number of true choices available to the individual for each decision are extremely limited. The experienced man has learned the options available to him. The production rate of the inexperienced man, however, is much lower as he must make his decisions from a larger array of choices. This array is larger because it includes many pseudo choices that can be ignored only with experience.

The industrial communication system is particularly well adapted to the application of automatic error reducing techniques, i.e., feedback controls. At all levels of the communication-production process it is normally possible to establish feedback channels which serve to compare the results of a communication with its original intent. Many of these feedback loops already exist in the factory under the headings of inspection, quality control, cost control, sales reports, production reports, and, finally, profit and loss statements. With accurate feedback, communication performance or error is usually obvious.* The failure of mechanical elements is readily traceable. Inadequate performance of the human elements may require a point by point application of the tenets of communication theory.

Efficient, accurate information flow is a product of two basic factors: channel capacity and code matching. Since all information above the channel capacity will be transmitted in error, the system capacity must be examined in detail, element by element. For evaluation of the human components, particularly in higher order decision making positions, the importance of experience, training, and long-time group association cannot be overemphasized. The intelligent man can learn by experience, but in the industrial situation, intelligence alone cannot replace experience. This is particularly true since human decision is based upon simulated or actual experiences. Within an inexperienced group or a group containing new and inexperienced members, communications will be very slow and redundancy will of necessity be high to preserve accuracy. Experiments in military group information transfer have shown that the conventional, so-called learning curve can be attributed to a combination of increased channel capacity and more efficient coding. As the individual learns his

---

* "Accurate Feedback" must be emphasized since the feedback channel is in itself subject to noise and error.

task, channel capacity increases with decreased semantic and psychological noise. Message redundancy can be materially reduced when the encoder is certain that the decoder has recognized the limited number and range of commands applicable to the situation. For example, in a newly formed radar warning group, an interpersonal message might be, "Corporal Watkins, 'Note and record the object showing in sector 32 of the radar screen and interrogate it for the code of the day.'" After months of association the same command may become simply, "Sector 32!" The encoder by then would know that Corporal Watkins was responsible for sectors 29, 30, 31, and 32. The command, "Sector 32" would automatically limit Watkins to a scan of sector 32 with an interrogation of any objects sighted. Decision making on Watkins' part had also been simplified. He had only to identify and survey one of four sectors upon receipt of a suitable command. The code necessary for complete transfer of information in this simple instance need only be four numbers.

Case Examples

The young engineer will start his career supplying information to both lower and higher decision making positions in an organization. It is important that he understand how the information he generates and the manner in which he communicates this information affects the performance of others.

An interesting application of the concepts of Information Theory occurred recently when the manager of a substantial manufacturing operation realized that he was being overwhelmed by the flood of paperwork across his desk. Despite the efforts of an efficient secretary, he was unable to keep up with the decisions required of him. The staff consultant who was given the task of organizing the information flow to the executive decided to apply Information Theory rules to the situation. The manager stated that the primary function of his position was to make decisions regarding the allocation of men and resources to meet the requirements of the division. (These requirements were brought to him in the reports of his subordinates.)

In studying the contents of the paper flow across the desk of the executive, the consultant shortly realized that the signal to noise ratio of the letters and reports was low. It was obvious that a more efficient coding procedure was required. It was also apparent that the executive had the channel capacity (experience, ability, and time) to produce the required decisions. The papers to be handled were found to fall into two general classifications: reports on work in progress, and statements of problems requiring an ultimate decision.

To better utilize the executive's capacity, a simple coding procedure was put into effect. Each letter was classified into either a progress report or a request for a decision. The members of the staff writing reports which offered information feedback on existing projects were requested to append a short statement on the progress that had been anticipated so that actual and anticipated results could be compared.

Communiques involving new problems were now required to clearly state the alternative decisions available; the recommended solution, and the manpower and resources required to effect this solution. Problems were not to be presented without a recommended solution. If this were not possible, it was mandatory that an outline of the steps necessary to generate an answer be submitted.

From the standpoint of Information Theory the logic of the above procedure is, of course, rather elementary. The executive had a finite channel capacity and could transmit information accurately at a corresponding maximum rate. The amount of information involved in each decision is based upon the number of alternatives available. A decision involving a large number of alternatives would, therefore, require more time to clear a fixed capacity channel than a decision with few alternatives. When he received problems with a proposed solution, including the costs, the executive's decision time was drastically reduced. He now knew all the alternatives envisaged by his subordinates and no longer had to consider an array of pseudo choices. In addition, his information transmitting task was further reduced, for now he had only to decide to grant or withhold an appropriation in light of the soundness of the recommendation. The amount of information connected with each decision had been materially reduced, allowing a corresponding proportional increase in the rate at which he could reach decisions.

As an example of downward communications, a small West Coast electronics plant found that its production rate was very low and its rejection rates unusually high. The company was manufacturing short runs of fairly complex electronic computer assemblies. An investigation of the problem determined that the young engineers on the project were turning out fine, detailed blueprints as they had been taught in school. Unfortunately, the women on the wiring and assembly lines could not read blueprints. The bottle neck was broken when the engineers presented the information to be utilized by the factory in a set of simple pictorial sketches.

Effects of Psychological and Semantic Noise

The effects of psychological and semantic noise upon organizational communication channel capacity can too often be described only as catastrophic. Although a number of articles have been written describing the advantages and the problems of two-way communication, it is seldom recognized that in most instances accurate communication through the upward or feedback path is literally impossible. In order for an organization to operate at all, some downward communication capacity must be formally established and maintained. Upward feedback channels are not used as frequently and consequently contain considerable semantic noise. This becomes particularly evident in the case of new or inexperienced management which has not shared experiences with its subordinates. In a like manner, communication from subordinates upward is subjected to much psychological noise. Many cases have been observed where management's displeasure over unfavorable reports has gradually forced subordinates to relate only that which they thought management wanted to hear. In these instances, information being passed upward has become 50% to 75% inaccurate or incomplete. Accurate management feedback can exist only in formalized channels where top management has earned the confidence of its subordinates.

The total responsibility for successful communication must rest with the original message encoder. He must devise a code which will match the code characteristics of the channel in order to produce a maximum transfer of useful information. He must produce information at a rate and with sufficient redundancy to minimize error. Where feedback channels have been established, he must recognize the changing nature of the channel and continuously rearrange his coding to further maximize the information transfer. As in our radar screen example, the coding redundancy can be reduced as increased capacity is developed. Each succeeding message recoder in a long communication channel shares the responsibility from his station onward.

The encoder must further recognize when insufficient channel capacity exists and why it exists. Insufficient capacity in general can be the results of the mechanical characteristics of the systems, i.e., human response times, or of an inadequate signal to noise ratio. While it is usually difficult to improve the human mechanical system, human channel capacity can frequently be enlarged by increasing signal strength or by reducing attendant noise.

To reduce noise, it is necessary to recognize its presence and type, whether mechanical, semantic, or psychological. An example that comes to mind is the case of a California school system where a conscientious superintendent found that he and his staff were unable to establish any degree of satisfactory communication with intelligent and interested parents. He ultimately recognized a semantic problem. For while he was speaking in the native tongue of the parents, he was attempting to relate pupil reactions which dated back to the parents' forgotten childhood, or teacher problems which the parents had never experienced. To reduce the semantic noise and establish a satisfactory channel, the superintendent persuaded groups of the parents to come to the schools for two weeks. During this period, they were given many of the typical lessons presented to their youngsters, and were exposed to the problems of the teachers at the same time.

A further example of management communication capacity was that of a large western manufacturing concern where the chief industrial engineer was unable to satisfy the sales manager on the order release procedure. The sales manager had handled order releases when the company was smaller and could not be told that existing production levels precluded the use of their older methods. The problem was solved when the sales manager had shared the industrial engineer's experience by taking over the responsibility of order release for a few months.

chapter 3 | COMMUNICATION SYSTEMS

The problems of communication can be studied from three viewpoints:

1. The communicator who wishes to effect some specific result and consequently must code his message in such a way as to overcome the mechanical noise of the transmission system plus the semantic and psychological noise of the receiver.
2. The receiver who must decide first what information the message was supposed to convey and secondly what course of action to follow as a consequence. Information theory, of course, applies to what was said; while decision theory concerns itself with what to do.
3. The designer of a communication system (to be used by others) who through the use of information theory can resolve problems concerning channel capacity, coding, and redundancy.

The engineer is concerned with all three of the above. The first two are exercised frequently in everyday activities. The third is encountered by engineers in a variety of guises ranging from deep space satellite communications to the organization of an industrial production department.

On the other hand, communication systems can be classified in yet another way with three different major divisions:

1. Man-Man Communication Systems.
2. Man-Machine Communication Systems.
3. Machine-Machine Communication Systems.

MAN-MAN COMMUNICATION

Interpersonal communications where one individual affects the behavior and activities of another certainly constitute the bulk of the communication activities of interest to us.* Looking at man as a receiver we see that all information must be taken into the brain through one or more of the five human senses. Therefore, if one mind wishes to influence another, there must be an intervening mechanical system which will perturbate the receiver's senses. (It is of interest that information can be transmitted *only* by changing energy levels in the transmitting and receiving systems.) The five human senses are, of course:

1. Taste
2. Smell
3. Touch
4. Hearing
5. Sight

The senses have been listed in their reverse order of usefulness in interpersonal communications. Some may question the order of the last two, but none question their importance. The problem of educating Helen Keller stemmed from her loss of both sight and hearing at birth. All of her information inputs had to be received through her taste, smell, and touch senses with the bulk of the input through touch.

Very little interpersonal technical information is transmitted through *taste* or *smell*. However, man often queries nature for information and sometimes receives data in the form of an odor or taste. An overheated electrical system can be detected by a pungent odor. The means of storing information for the later actuation of these first two senses is also very limited and quite crude.

A useful but limited amount of technical data is transmitted interpersonally directly through *touch*. For the blind, braille forms an indispensable link with man's storehouse of knowledge. While the transmission rate is somewhat slower than that of oral or sight transmission, any information that can be expressed in words can be stored in braille and received through touch.

The student of college age who has now spent 10 to 12 years in classrooms does not have to be told that a goodly percentage of interpersonal communications are attempted through the media of *sound*. In fact, our civilization has been likened to a modern Tower of Babel. To exercise the audio channels of the receiver, an acoustical system must exist which couples the transmitter to the receiver. In addition to the acoustical coupling, we often find many other intermediate stages including electrical, electromagnetic and mechanical links in the overall communication channel.

---

* We recognize that most other living creatures have means of communication which are vital to their existence.

At this point the reader should refer back to the general description of a communication system (Figure 2). All of the elements described are now present. For oral communication, certainly the mechanical noise level of the channel is important. The presence of semantic and psychological noise can be determined with relative ease. The coding and redundancy problems are always with us and must be solved if the desired accuracy is to be achieved within the allowable time.

Although oral communication has always been important, its usefulness has been greatly extended within the past 30 years. During this time the devices for the permanent storage of information with audio read-in and audio read-out have been brought to a high degree of perfection. Sound waves can not only be transmitted over long distances at low cost, but they can also be recorded for future play back at very nominal expense. This means that oral instructions are not necessarily lost within a few seconds but can be mechanically repeated as often as desired. As a consequence, recorders are replacing visual instruction in many instances where sound is more effective than sight. In modern aircraft, the pilot's instrument panel has become so complicated that a warning light could go unnoticed. Recorders are being used to give the pilot a spoken warning of dangerous conditions. In the factory audio assembly instructions often replace blueprints.

In spite of the increase within recent years of information stored for audio read-out, most of man's knowledge at the present time is stored in forms suited for *visual* read-out. Storage of information in pictorial or written form goes back to prehistoric times. Certainly there are more techniques available for visual communication and information storage than for any or all of the other four senses.

The transmitters of information for visual read-out include:

1. Smoke signals
2. Gestures, including sign language
3. Semaphore
4. Lights (remember Paul Revere)
5. Color codes (used extensively in electrical manufacturing)
6. Photographs
7. Motion pictures and television
8. Pictures (stationary) and sketches
9. Blueprints—mechanical drawings
10. Hieroglyphics and mathematical symbols
11. Printed words
12. Handwriting
13. Painting

Modern technology is rapidly extending our resources for the storage and transmission of information. However, since we have little hope of changing man himself during the next few generations, it is still highly probable that most of the information impinging upon the brain of the engineer will be received through his most efficient sensor—the eye. The engineer will not only have to master the older and more conventional interpersonal communication skills of writing, drawing, and symbolic representation, but he also will have to become familiar with the advantages to be gained from machines, including motion pictures and television.

The choice of transmission channel and code will again depend upon the characteristics of the information to be transmitted and the nature of the receiver. Consider only the problem facing a project engineer who has designed a complicated machine. To communicate with the president of the corporation, he might produce a carefully organized formal report couched in nontechnical terms. For the president, he would delimit the problem, outline the sales potential of the product, and perhaps advance a manufacturing budget. On the other hand, for his immediate engineering superior, the engineer would produce an entirely different report. The technical aspects of the problem would be fully explored. Computations might be included. Test procedures and test results would be described.

For the manufacturing departments, the communication will often be reduced to a detailed set of blueprints. To direct skilled and semi-skilled assembly personnel, a range of instructions could be available from pictorial sketches carefully broken down into step by step operations up to expensive motion pictures.

MAN-MACHINE COMMUNICATION

The thought of man talking to machines or of machines directing the activities of man is perhaps somewhat repugnant to most of us. However, this situation has been with us for many years. One has only to stand at a busy intersection to see tons of machinery and hundreds of people being directed by the flicker of a 150-watt bulb.

The engineer will frequently query nature. We call such activities testing. In these cases man's senses are greatly extended through the use of instruments and machines. Testing requires a communication system between the phenomena investigated and a human sensor.

The problems of man-to-machine communication are particularly interesting. A major facet is that of proper coding. In recent years much effort has been expended on the problem of machine languages to allow man to more efficiently communicate with modern high-speed computers.

In the industrial organization, the engineer is constantly directing the activities of the production machinery. In the past most of these instructions were carried out through the intermediate efforts of other men. The machine tool would be operated by a machinist who would interpret the blueprint produced by the engineer. Today with increasing frequency the intermediate steps are being eliminated. Machine tools are in production with fully automatic numerical controls. The engineer bypasses both the machinist and the blueprint. He communicates directly with the machine through the media of a punched or magnetic tape which has been produced in the engineering department in lieu of a mechanical drawing.

It is not the intent of this book to explore man-machine or machine-machine communications any further. For the sake of completeness, however, the student must realize the full extent of the communication systems with which he will become involved.

COMMUNICATION SKILLS

The latter chapters of this book are devoted to the basic interpersonal communication skills of writing, speaking, and drawing. Too often, the engineering student is told that he needs English composition, report writing, sketching, or drafting because "they are good for him." The implication, unfortunately, is that development of these skills will make him a better man and obviously a better citizen. These considerations can be important. It is certainly desirable to have the engineering profession represented by men who speak grammatically and who can produce reports that are both coherent and reasonably well punctuated. However, these needs are secondary compared to the simple and basic communication requirements that the engineer must satisfy. All engineering is dependent upon the accurate, economical, and rapid transmission and processing of information. The rules of English grammar and punctuation form a useful common code. Ungrammatical sentence structure and poor punctuation introduce semantic noise. This costs the reader time and enhances the possibility of error.

Every report and every sketch that the engineer produces must be considered as part of a larger communication system. The information to effect the desired result must be coded in light of the characteristics of the channel and of the receiver. Enough redundancy must be introduced to deliver the desired accuracy. Yet the code must not be so complicated as to render the system uneconomical. As a very simple example, consider the problem of a sketch in which a shaft is to be shown. If the engineer is communicating with himself or with another engineer, he can often depict the shaft by a center line and a dimension. For unskilled shop personnel, much more redundacy is required. The shaft might be even drawn in complete perspective with shading to emphasize its shape. The difference in time required to produce the two representations is considerable. In the first case the redundancy is wasted. In the second case, it is essential to insure understanding and accuracy.

In playing the communication game, the engineer is unusually fortunate. With information theory he possesses a complete system for the evaluation of each communication problem. The elements to be determined are:

1. The desired result
2. The nature of the receiver
3. The nature of the transmitter
4. The mechanical noise of the system
5. The semantic noise of the system
6. The psychological noise of the system
7. The resultant overall channel capacity
8. An effective code considering the role of redundancy and the rationale of standards.

The remaining chapters will relate the fundamentals of the communications skills to these basic requirements for effective communications.

REFERENCES

1. S. S. Stevens, "Introduction: A Definition of Communication," *J. Acoust. Soc. Am.*, **22**, 6 (1950) 689 ff.

2. Stanford Goldman, *Information Theory*, Englewood Cliffs, N. J.: Prentice-Hall, Inc., 1953.

3. Claude Shannon and Warren Weaver, *The Mathematical Theory of Communication*, Urbana, Illinois: The University of Illinois Press, 1949.

4. L. Marton, ed., *Advances in Electronics*, New York: Academic Press, Inc., 1951.

5. W. G. Tuller, Information Theory Applied to System Design," *Trans. A.I.E.E.*, **69** (1950).

6. *Current Trends in Information Theory*, Pittsburgh: University of Pittsburgh Press, 1953.

7. Jergen Ruesch and Gregory Bateson, *Communication the Social Matrix of Psychiatry*, New York: W. W. Norton & Company, Inc., 1951.

8. Homer Jacobson, "Information Theory and Life," *Am. Scientist*, **43**, 1 (1955) 119.
9. Otto J. M. Smith, "Economic Analogs," *Proc. Inst. Radio Eng.*, **41**, 10 (1953) 1514 ff.
10. Martin Joos, "Description of Language Design," *J. Acoust. Soc. Am.*, **22**, 6 (1950) 701 ff.
11. Oliver H. Strauss, "The Relation of Phonetics and Linguistics to Communication Theory," *J. Acoust. Soc. Am.*, **22**, 6 (1950) 709 ff.
12. S. S. Stevens, ed., *Handbook of Experimental Psychology*, New York: John Wiley & Sons, Inc., 1951.
13. Eugene L. Hartley and Ruth E. Hartley, *Fundamentals of Social Psychology*, New York: Alfred A. Knopf, Inc., 1952.
14. Allen B. Rosenstein, "The Industrial Engineering Application of Communication-Information Theory," *J. Ind. Eng.* **6,** 5 (1955).
15. J. Pierce and J. Karlin, "Reading Rates and the Information Rate of a Human Channel," *Inst. Radio Eng. WESCON Conv. Record*, Pt. 2 (1957) 60.
16. George C. Sziklai, "Some Studies in the Speed of Visual Perception," *Inst. Radio Eng. Trans. Inform. Theory*, Sept. 1956, 125 ff.
17. H. Quastler, *Information Theory in Psychology*, New York: Free Press of Glencoe, Inc., 1955.
18. P. M. Fitts, "The Information Capacity of the Human Motor System in Controlling the Amplitude of Movement," *J. Exptl. Psych.*, **48** (1954) 381 ff.
19. P. M. Fitts and P. L. Deininger, "S-R Compatibility Correspondence Among Paired Elements Within Stimulus and Response Codes," *J. Exptl. Psych.*, **48** (1954) 483 ff.

part
II

COMMUNICATION
PRACTICE

chapter 4

THE READER

IMPORTANCE OF FEEDBACK

One of the most serious difficulties associated with writing as a communication system is that there is no built-in feedback channel from the destination to the source, i.e., from the mind of the intended reader to the mind of the writer. In oral communication we often receive direct feedback from our listener, both verbally and physically. We use this feedback to adjust our transmission to improve the accuracy of our communication. From his questions we can re-evaluate his channel capacity, modify the coding system we are using, and correct any mismatch in semantics. If he appears puzzled, we introduce additional redundancy into the signal by reiterating or by going into further detail. In the writing process, however, we receive no reaction from our reader during the critical state of encoding our communication. He cannot say, "Hold on a minute, I still don't understand your analysis in Section 3," or, "How about a little more background on this project before you propose a solution?" Therefore, we are obliged to anticipate our reader's questions and to supply answers at the exact places in the report where they will be needed. In terms of communication theory, the report writer must determine what type channel will be best suited to his reader and then select a code that will match the code characteristics of the channel, strengthening the signal whenever the reader might encounter noise.

IDENTIFYING THE READER

The first step in preparing to write a technical communication is to identify the reader since he represents the destination in the communication system. Just who is to be "the reader?" What position does he hold?

What is his background and experience? How much does he already know about the subject? How extensive is his technical and general vocabulary? What use will he wish to make of the report? Only when the writer can answer these questions with some measure of certainty will he be ready to begin his report.

"The reader," of course, may be more than one person. When this is the case, each receiver must be classified according to his importance *as a reader*. Whenever a report is to form the basis of a decision or whenever action of any sort is to be taken on the information the report carries, the person or group who will make the decision or take the action is the *primary* reader. Others who may receive the report for information only are the *secondary* readers, regardless of the company positions they hold.

It is important to make this distinction in rank among readers because the writer should always cater to the needs of the primary reader first. This is the best way to guarantee a wise selection of information, channel, and code for transmitting a message to those who "need to know."

Once the writer has determined how to meet his responsibilities to the primary reader, he should then see if he can satisfy at least some of the needs of the secondary reader. Common techniques for achieving this objective are:

1. Include material of interest to the secondary reader, but relegate it to an appendix.
2. Present background material with which the primary reader is familiar in a letter of transmittal or a foreword, addressed specifically to the secondary reader.
3. Divide the report into two parts. Use the first part for the primary material, the second part for the secondary. For example: Part I—An Economic Appraisal of the Project, Part II—The Engineering of the Project.

The above example assumes that management is the primary reader and that only one report is being written. If engineering also were primary, then it is likely that two separate reports would be in order. The two-report procedure, however, is time consuming and should be followed only under these conditions: when there is such a dissimilarity in the background or interests of the readers that one report would be unsatisfactory; or when the volume of the material is so great as to make a single report cumbersome.

If you ever are given a report assignment and have no idea who will be reading your writing (and can not find out), you could do worse than to address your report to a hypothetical reader who has the same background and interest as you have but who is not informed about your particular project.

ANALYZING THE READER'S CHANNEL CAPACITY

All readers, whether primary or secondary, have certain channel-capacity characteristics in common. They also have special channel characteristics, some of which are psychological and thus differing greatly from individual to individual. Only the common channel characteristics can be predicted and effective ways to meet them prescribed; the individual characteristics have to be appraised with each new reporting situation. The remainder of this chapter, therefore, will concern itself with the common characteristics of the human communication channel.

If we put ourselves in the reader's place, we can readily see that there are two important factors which should influence our writing:

The first is that the reader is at the beginning of the investigation, whereas the writer is at the end. The writer is usually reporting on a piece of work done—his project is finished and now he is recording procedures, results, conclusions, and recommendations. The reader, however, is usually starting new on the investigation.

The second is that the reader is intelligent, but uninformed. He is ignorant only in the sense that he does not have the information the writer possesses and thus needs to have things explained to him more patiently and simply than the writer supposes.

Once we've examined the position of the reader, it is clear that he has certain important basic needs. First, he must be able to get the intended message clearly and without unnecessary work. As readers, you know how often you must puzzle out a piece of writing to get the intended message. In a piece of writing of this nature you can be sure the writer has not thought of the reader's needs very carefully.

Secondly, the reader wants to be able to read the communication rapidly. Scientific and engineering writing is read at work, in the laboratory, in the office, in places where the pressure and pace of activity force a rapid reading. The well-paced report will allow the reader to move through it rapidly; he must, if he is ever to finish the pile of papers on his desk.

Finally, all readers want to be able, if necessary, to read discontinuously. A reader is often interrupted during the reading of a report. The careful writer, realizing this, will provide convenient stopping and starting places (such as headings and subheadings) that will help the reader in this situation.

HOW TO CODE THE MESSAGE FOR MAXIMUM EFFECTIVENESS

Up to this point, we have examined the nature of the reader-writer relationship, identification of the reader, and the common characteristics of the reader's communication system. Now we will try to show some of the ways reader needs can be satisfied. The first and perhaps most obvious way is in the subject matter itself.

1. *Make the title meaningful and, if possible, brief.* Single-word titles are satisfactory only for a piece of writing that covers a broad or general topic. These titles can be misleading when the subject matter is limited to a segment or a specific area of a larger topic. For example, the writer who has limited his subject to a single, operating condition of a specific type of subminiature triode must compromise on brevity for the sake of preciseness in his title. In the title, the reader needs an accurate statement of what the piece of writing is about. This increases the reader's channel capacity for it immediately eliminates many of the pseudo possibilities and directs the readers toward the true subject material (see page 27 of Part I).

2. *Summarize the high points.* One of the best ways you can bring the important points of your investigation to the attention of your readers is in an informative summary or abstract. Capture the reader's attention immediately by placing the main conclusion up front. Keep the abstract short, but do not hesitate to be quantitative. This again increases the reader's channel capacity by eliminating many pseudo possibilities. The reader now knows your conclusions. He will not waste time while he reads trying to question what you are driving toward. He knows in advance and can thus concentrate on the steps in your logic.

3. *Provide sufficient background material.* The amount of background material in any report should be directly related to the intended reader's knowledge of the investigation. For example, the writer's immediate supervisor might not need any background information at all, since he may be thoroughly familiar with the project. On the other hand, if the report is addressed to a reader who is not familiar with the project, the writer must take particular care to fill that reader in on the "why" of the investigation. Sometimes the writer is addressing both groups, however. Then the amount of

background material is dictated by the reader group with the least knowledge of the investigation. The informed readers can always skip over what they already know. This step is taken to reduce the semantic noise in the message. The reader must be given enough background to "understand" the local situation.

4. *Associate the unfamiliar with the familiar.* In the following excerpt from a report on new developments in pulsed-circuit test equipment, the writer carefully sets the stage for his description of the new developments by bringing in a familiar concept and going on from there:

"The term 'test equipment,' when used to describe electronic devices, usually brings to mind signal generators, oscilloscopes, vacuum-tube voltmeters, and other commercial measuring instruments. These units are excellent for testing communication systems where a single input produces a single output and the intelligence is provided by some method of modulation.

"But in a large-scale system of pulsed circuits, such as a digital computer, hundreds of signal lines must be switched to form pulse channels, and the usual method of transmitting intelligence is to supply a pulse on a particular line at a specified time. This problem of pulse routing, plus the fact that the pulse must meet a required amplitude, shape, and duration, makes system testing with conventional test equipment extremely difficult.

"The test units described in this paper are special devices, each designed to perform a specific function in pulse circuitry. . . ."

Again we improve the accuracy of our transmission if we reduce semantic noise by defining new terms or old terms applied in new or unusual ways.

5. *Describe the whole before the parts.* Just as the reader needs a picture of the whole problem in the introduction, so he needs similar guides throughout the presentation of evidence. In order for him to understand a new concept, a new machine, or a new method, he must first have a clear idea of the essence, function, and purpose of the concept, machine, or method—considered as a whole. He needs an over-all framework into which he can fit the parts as they are described. For example, if the reader is first given a general description of a chemical process, he will be ready to tackle the details of the many steps in the process. The verbal picture will help him in much the same way that the illustration on the box helps the man about to assemble a jigsaw puzzle.

Here again we increase reader capacity by eliminating in advance the pseudo choices which may arise in his mind regarding the role of the parts and their relation to the whole.

6. *Develop a sound structure.* Once the reader has a picture of the whole, the parts should be presented to him in a pattern that is consistent with the nature of the subject matter. There are many ways of unfolding a story, e.g., development based on time, classification, order of importance, cause and effect, or logical sequence. Usually, one method is better than another for a particular situation. It is up to the writer to find it, and until he gains experience, cut and try may be his only approach. For example, the construction of a device might be described by working from the inside to the outside, from the top to the bottom, or from the left to the right. If the general description does not immediately suggest which method to use, the writer should try several and compare. The one that was easiest to write should be the one easiest to read.

Our problem here is that of the compatibility of the ensemble of stimulus and response symbols—in other words, the fit of the encoding and decoding processes as described in Chapter 2, page 21.

7. *Emphasize the primary ideas.* All the details that go into a description will not be of equal importance. Yet sometimes the reader inadvertently attaches undue weight to a statement simply because the writer was not careful in separating and labeling his primary and secondary evidence. Then, too, readers feel that the location and the amount of space allotted to a topic have a direct connection with the importance of that topic. If you find that a certain piece of secondary information contains so many details or qualifications that it requires a large amount of text, you should summarize the information and relegate the details to an appendix. Descriptions of procedure often fall into this category, as do development of equations and discussions of test equipment.

The reader's communication channel is thus free to process the important ideas. Valuable capacity is not to be wasted on secondary information.

8. *Separate fact and opinion.* Naturally, every reader wishes to be able to distinguish between fact and opinion and between the views of the author and those of others. Thus every statement of opinion should be labeled and the person responsible for each identified. Plagiarism is not the issue. But since the pronoun "I" has been outlawed from many areas of technical writing, the author's personal views are frequently mistaken for those of accepted authorities in

the field. Such common expressions as "It is believed that . . ." and "It is concluded that . . ." are cases in point. If personal opinion is *not* in order, then any way of expressing it is wrong. However, if the reader wants the author's views, then "I believe" is better than "The author believes," "It is the author's belief that," "It is believed that," or "We believe." For the formal company report, though, the author must say "The Engineering Department recommends," "B W & N believes," etc., since the organization, not the individual, is held responsible for what is said under its letterhead.

Two important communication principles are involved. If the reader is misled into accepting opinion for fact, then the accuracy of the communication has been decreased and more noise introduced. On the other hand, the reader may lose channel capacity if he has to continually guess whether the material represents fact or opinion.

9. *Use precise, straightforward language.* For technical writing, the best style is one that does its work quietly in the background without calling attention to itself. Clarity and efficiency of expression are needed, not impressive language. We offer here, then, a simple plea to minimize the semantic noise of the system: match your output to the decoding ability of the reader. Some of the major sources of semantic noise, apart from mismatches in vocabulary, are illustrated below.

Fuzzy qualifiers—	Plates of *appreciable* thickness. (How thick?)
	A *relatively* high temperature. (How high?)
	A *small number* of failures. (How many?)
Euphemisms—	Six tests were run and the firing curves were very smooth for all except the first, third, fourth, and sixth.
Overformal words—	Herewith are enclosed the requisite documents employees are requested to submit subsequent to the termination of their period of probation.
Jargon and coined words—	The system can be introduced with *effectivity* within six months.
	All the components are *ruggedized*.
Clichés—	Last but not least, we intend, in the long run, to explore every avenue which might lead us to a solution along this line.

Deadwood— In the event that—(if)
In view of the fact that—(because)
Despite the fact that—(though)

HOW TO CODE A MESSAGE FOR THE RAPID READER

Today every reader is in a hurry. He has so much to read that he has to skim everything if he is to get through the daily pile of papers on his desk. As mentioned earlier, he may also have to read discontinuously, since the pressures of his job seldom permit him to read a report from cover to cover in one sitting.

From a practical viewpoint, the writer must cater to these needs. He must build ease of reading into his style and format. Actually, the job is not so difficult as it is time consuming. But this is justifiable, since the object is to cut down on reading time.

The following are suggestions for tailoring the prose and mechanics to the rapid reader.

1. *Use descriptive headings and subheadings freely.* One of the most effective ways to facilitate rapid reading in a technical report is through the use of headings. Because of their important function, you should never be afraid to insert a heading when you feel one would help the reader. Descriptive headings and subheadings act as an internal table of contents, quickly directing and orienting the reader to the various topics covered in the report and permitting him to read selectively. Also, headings help the reader quickly find his place after he has been interrupted in his reading.

2. *Place topic sentences at the beginning of paragraphs.* Placing the topic sentence at or near the beginning of a paragraph allows your reader to skim through your report. Although he may miss detailed development of main ideas, he still will have a good general idea of what you say in the report. Also, for the man who will read your report completely, the topic sentence at the beginning of the paragraph will serve as "the whole before the parts"—the main idea before the details. This form of controlled redundancy greatly improves the accuracy of the transmission.

3. *Use simple sentence structure when the thought is complex.* Whenever the thought is involved or otherwise difficult to describe, the sentence structure should be simple. For example, three short sentences on a complex topic are easier to read than one long one because the reader has to deal with only one idea at a time. This prin-

ciple, as illustrated below, demonstrates how semantic noise is greatly reduced with simple sentence structure.

Original

An increase in the carbon content makes the steel harder, though more brittle, and machine parts, such as gears, which need a hard surface to resist wear and a ductile interior to stand up to sudden shocks without breaking, are given these properties by the process of case-hardening.

Suggested Revision

An increase in the carbon content makes the steel harder, though more brittle. However, machine parts, such as gears, need a hard surface to resist wear and a ductile interior to stand up to sudden shock. These properties can be given to steel through the process of case-hardening.

4. *Make full use of graphic aids.* Curves and tables that summarize detailed results form an invaluable source of desirable redundancy. Most readers have trained themselves to extract the information they need, at a glance. But always supply captions and legends and refer to the figure at that point in the text where the reader needs the information. Ideally, graphic aids should be placed immediately following their reference in text. (Chapters 10 and 11 treat this subject in detail.)
5. *Adjust the pace to fit the subject and the reader.* If the pace of information flow is too rapid (that is, if the reader's mind cannot keep up with his eye), the reader will have to backtrack and the efficiency of the communication drops. If the pace is too slow, the reader will try to second-guess the writer and consequently may miss important bits of information. Technical description requires a slower pace than does technical narration because narration has a sort of "built-in" redundancy. A common way to control pace is through the length of sentences and paragraphs.
6. *Avoid footnoting secondary or reference material.* If secondary or reference material bears directly on the topic under discussion, try to work this information parenthetically into the text. Footnoting forces the reader to drop his eyes from the text to the footnote and back to the text again—a tiring process that badly hinders rapid reading. Secondary or reference material not bearing directly on the discussion should either be relegated to the appendix or omitted.
7. *Provide white space around the text.* The human eye can process only a finite amount of information per square inch of page. Even in a well-written and well-organized technical report, the reader can be slowed down by the needless crowding of information on a page. Give the reader wide margins, pronounced paragraph indentations,

space between headings and text, short paragraphs, and double space between lines whenever possible. An uncluttered format reduces reader fatigue and promotes rapid reading.

SUMMARY

A technical report has a reason to exist only if it has the effect the author intended upon the reader. Consequently, all of the principles of sound reporting are based upon reader information processes and decoding needs. Every decision a technical writer makes—whether on content, style, or format—must be justified on the grounds that the result will help the reader; each new reporting situation demands a fresh appraisal of the potential audience. Whether we like it or not, the man at the receiving end is the important man.

chapter 5

THE WRITER AND THE REPORT

As a project engineer involved in engineering design you will have to write numerous technical reports. Among the more common types will be:

Specifications	(which *others* may have to bid upon)
Design Proposals	(to tell *others* how *you* would design a system)
Operating Instructions	(to tell *others* how to operate a device)
Manufacturing Instructions	(to tell *others* how to manufacture a device)
Test Procedures	(to tell *others* how to test a product)

In addition, you will be responsible for reporting the status of your project through periodic progress reports, and sometimes for documenting the entire investigation through a final or summary report.

WHY WRITE A REPORT?

As we pointed out in Chapter 3, writing is but one of the transmitters that an engineer might use to convey information and ideas to others. As a matter of fact, on a day-to-day basis it is used much less frequently than oral transmission and certainly no more frequently than other systems designed for visual read-out, e.g., drawings and photographs.

Why write reports, then? The answer lies partly in convention, partly in economics, and partly in the fact that the channel characteristics of writing, taken *in toto*, are unique.

Engineers have been writing reports for centuries. Through this mechanism the literature of the profession has been established; through it much of the literature of the near future will be accumulated. Conventions do not disappear over night—even though they may diminish in usefulness.

If convention were the only attribute to support writing, however, it would not be worth our time nor yours to discuss report writing here. Economics plays an important role. Coding and decoding signals on film is very expensive compared with writing on paper; television also is a very costly medium. Naturally, there are more ways than one to look at costs, but by and large, reports cost less than their more glamorous counterparts. All of the various transmitters, whether designed for visual read-out, for audio read-out, or for both, have certain channel characteristics that set them apart. Selection usually is made first on the type of read-out desired; then from the available transmitters within that type, weighing cost against whatever performance variables are considered critical to the immediate communication.

In addition to low cost, the important characteristics of writing that might prompt its being selected from among other visual methods are these:

1. It is readily available.
2. It requires only simple tools for encoding, transmitting, and decoding.
3. It offers a large variety of symbols that, presumably, are familiar to both encoder and decoder.
4. It uses a basic coding system also presumably familiar to both encoder and decoder.
5. It provides for control of redundancy.
6. It is a convenient, interpersonal communication system; i.e., single source to single destination.
7. It can easily be combined with other means of visual read-out, such as sketches, diagrams, and graphs.
8. It is adaptable to both short and long messages.
9. It provides a simple, economical means of information storage and mass transmission.
10. It is also a versatile system with large channel capacity. The information it carries may treat any technical subject: the operation of a machine, the behavior of a circuit, the development of a theory, the design of a system. It may follow a variety of expository forms: it may instruct or command; it may define; it may analyze; it may persuade; it may compare; it may summarize. Writing can also be tailored to meet the needs of different receivers, as we already have pointed out in Chapter 4.

Before we continue with our analysis, however, we must reiterate that other transmitters do indeed possess some of the above characteristics;

furthermore, they have individual characteristics not attributable to writing. It is the total combination (characteristics and cost factor) that keeps the writing transmitter in business.

WRITING AS A COMMUNICATION SYSTEM

We are now ready to look more closely at the elements of a complete communication system (Figure 2, p. 21) as they apply to report writing.

The *Information Source* is the mind of the writer (assisted by external memory devices such as notes and data sheets). It selects the message to be transmitted and determines the intent.

The *Semantic Encoder* performs the mental actions of selecting the channel to be used and encoding the information into appropriate symbols (words, numbers, mathematical symbols, etc.).

The *Transmitter* performs the physical act of putting the symbols onto paper as signals.

The *Mechanical Channel* is the written report.

The *Receiver* performs the physical act of reading the signals as symbols.

The *Semantic Decoder* translates the symbols into a message.

The *Destination* is the mind of the reader, serving as a decision center to be influenced by the message.

Noise: *Semantic Noise* includes not only faulty word choice but also improper sentence structure, failure to put words into a meaningful context for the reader, and poor organization of thought.

Mechanical Noise includes errors and inconsistencies in typography, poor physical layout of illustrations, overcrowding of text, poor reproduction qualities, and even such items as a binding that prevents the reader from keeping the report open at a given page.

Psychological Noise is any emotional reaction by the reader that decreases his channel capacity, thus distorting the message. Doubt, disagreement, boredom, and even anger are the common negative reactions of technical readers. The source of psychological noise may be the message itself, semantic or mechanical noise in the message, or some external stimulus.

In a simple communication situation consisting of a writer, a handwritten report, and a reader, the elements of the system line up as follows:

   The Writer   The Information Source
             The Encoder
             The Transmitter

The Report   The Mechanical Channel
The Reader   The Receiver
                  The Decoder
                  The Destination

Because of its simplicity, this system has the potential of being as free of noise as is possible to achieve in a written communication.

In a more involved communication situation, the noise probability increases. Here are five cases; the new elements are in italics.

1. Writer—*Secretary*—Report—Reader
2. Writer—*Editor*—Secretary—Report—Reader
3. *Co-writers*—Editor—Secretary—Report—Reader
4. Co-writers—Editor—Secretary—Report—*Multiple Readers*
5. Engineer (doer)—*Technical Writer*—Editor—Secretary—Report—Multiple Readers

In each case the new element is a human component. In Cases 1, 2, 3, and 5, this component is added to the input; in Case 4, to the output. To simplify the illustrations, we have assumed the information to be transmitted, the intent of the originator, and the mechanical channel as being constant.

Although the addition of human elements increases the probability of noise occurring, luckily the special training of a secretary, an editor, and a technical writer permits us to assume that these people can remove more noise than they create. We should not overlook the basic fact, however, that the more people involved in producing a piece of writing, the greater the chance of error. This is all too true in the case of co-writers (Case 3) and multiple readers (Case 4). In Case 5, the problem arises from the need for a double transmission—one from the engineer to the writer and one from the writer to the reader. Perhaps not so obvious, but equally dangerous, is the case of multiple editors (not shown). Here the amount of semantic noise produced can become overwhelming.

CHARACTERISTICS OF THE MECHANICAL CHANNEL

The mechanical channel consists of the code characteristics of the language it carries, the physical characteristics of the report itself, and the organizational structure of the subject matter. Since the problems of language are discussed in other sections and the physical structure varies considerably from one report to another, we will examine only organizational structure here.

Every engineering report has at least two functional parts: an introductory section and an evidence section. Many have a third or terminal section. In the above sequence, these parts satisfy a psychological need of the reader: he is briefed on the subject before he is asked to examine it; he examines it before he is asked to evaluate it. The logical overlapping of information by carefully controlled redundancy from one section to the next makes this sequence a powerful communication tool.

Good communication practice and convention have established the introduction-evidence-evaluation sequence as standard coding for technical reports. However, this does not mean that the standard must be followed each time. Any writer has the right to change the order at any time, provided he is able to justify the change on the grounds that it will help the reader. For example, a writer might place his conclusions and recommendations up front instead of at the end because he has learned that the reader is primarily interested in evaluation.

Most of the time, a change in the standard organizational structure will void or at least weaken the advantage provided by redundancy. The writer should make up for this loss by inserting new redundant elements. In the example above, the conclusions and recommendations were placed ahead of the introductory material. To make up for the loss of redundancy, the writer might insert a brief foreword between the title and the conclusions and recommendations.

PROBLEMS AT THE INFORMATION SOURCE—
PREPARATION FOR WRITING

Before you as a writer can determine what message is to be transmitted and what effect you wish it to have on the reader, you have several tasks to perform:

1. Analyze the information you have gathered.
2. Identify the reader you wish to communicate with.
3. Filter and synthesize the information in your own mind.

Since we already have discussed the reader in Chapter 4, we will treat only the other two points now.

As an engineer you will be continually faced with solving problems during a design project. Whenever you develop the alternatives from which to make a choice or form a judgment you are dealing with information. Thus, you start to produce the raw materials of a communication from the moment you begin the design process. How well you will be able to assimilate all the information that has collected at the final stage of the process depends largely upon how efficiently you have stored and organized it.

Project Notebook

The external memory device most widely used by engineers is the project notebook. You and your notebook form an intimate communication system; no communication will be achieved, however, if you do not code your input properly. Do not write as if for a diary; narrative is easy to write but it is an inefficient code for your purpose. Since you will have to weigh, rearrange, select, and reject bits of information before you begin to write a report, code your notebook entries accordingly:

Use main headings and subheadings.

Separate physically the important from the trivial.

Label key facts and ideas.

Cross-reference your material.

Periodically (once a week is reasonable), summarize the high points and evaluate the progress.

Project Folder

In a group project, each member must be kept informed of the plans and progress of his colleagues. The project folder is an extension of the project notebook. In effect, it serves as a central file for all information bearing on the conduct of the investigation. Each member of the group makes a carbon of any communication he initiates, and files the copy in a predetermined division of the folder. In the case of the project notebook, only copies of the summaries with pertinent drawings would be included. The folder usually is kept in the group leader's office and is available to any member of the group at any time. It can be particularly helpful during a group conference.

Card Notes on Literature Search

If your project involves a literature search, you might find this technique for note-taking handy. It consists of a master index for bibliographical data and individual cards for each topic you wish to record. The standard 3×5 inch lined cards are satisfactory, although you may use a larger size if your entries will be lengthy.

This is how the system operates:

1. Prepare a bibliographical index of every book, article, report, etc., you examine. Assign a reference number to each.
2. Write your notes on separate cards, one for each topic. Put topical headings, index reference numbers, and page numbers on all cards. Indicate direct quotes by using quotation marks.
3. After you finish all the readings, group cards with the same or similar topical headings. You now have ready for analysis and evaluation everything you read on each topic.

The card system also has these advantages over a notebook: you can spread the cards out and make quick comparisons; you can rearrange the cards at will; you can use them as notes for an oral report; and you can add new cards to the deck without having to make awkward inserts. A sample card is shown below.

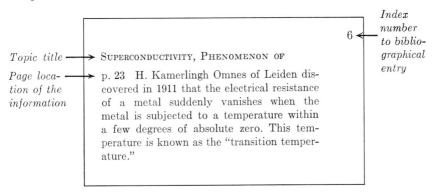

CARD NOTE ON LITERATURE SEARCH

Statement of Thesis

Before your mind transfers its thoughts from memory to encoder you will wish to synthesize the raw information you have collected, so that you may establish the intent of your communication and thus the emphasis of your message. You accomplish this synthesis by creating a statement of thesis—a statement based on the information you have about your project and the information you have about your reader.

Suppose you had just designed a piece of test equipment and are told that Mr. X of the Jones Company is looking for a device to test his quality control equipment. You investigate his problem and write him a short report proposing that he use your equipment. Before you write the report you mull over what you want to say:

"Mr. X has a testing problem."
"I have new test equipment."

"My equipment will solve his problem."
"He should use it."

In simplified form, the above constitutes a statement of thesis: it reveals the emphasis and the intent. For more involved reporting situations, however, the statement may not be this obvious nor so quickly arrived at. But you would use the same general procedure.

A Practical Outline

A practical outline is one that will help you coordinate your functions as information source and encoder. It serves as an excellent memory device if you are willing to invest more than token effort in preparing it.

Since the bulk of your report will contain the body of evidence you have collected, spend most of your outline time on these two important operations: filter the raw material of your evidence through a statement of thesis, and carry the outline to at least a level of detail that provides an entry for each paragraph of your proposed report.

The Filtering Operation

Determine exactly what results you want to have achieved when the reader finishes reading your report. What general conclusion should he draw from the evidence? This is your central theme or thesis and reflects your intent. Before you begin to outline, try to state the thesis in a sentence or two. If you're not satisfied, re-examine your evidence and try again. When you believe you have interpreted the significance of your investigation correctly, filter your notes, data, observations, results, etc., by weighing each entity against the statement of thesis. Rate your material as either primary, secondary, or irrelevant. If you are rigorous in forming your thesis and in applying it to your evidence, you will know exactly what to include in your outline and what to stress.

The Paragraph Technique

It's important to have entries in your outline that stand for paragraphs in your report because paragraphs are the building blocks of a written communication. They are the basic units in which information is transmitted to the reader; therefore, they must be represented if the outline is to be of real use.

The procedure for representing paragraphs is not difficult. Paragraphs are developed from topic sentences, and in technical writing a good paragraph announces its topic at or very near the beginning. All you have to do is to write a notation for the topic sentence and then list under it the points you wish to bring out. The number of points will indicate the potential length of the paragraph. (If the number is high you may need to make two paragraphs.)

PROBLEMS AT THE ENCODER—WRITING

Practically all writers have some trouble converting thought symbols into written words that accurately represent the intended message. The major source of their problem is the semantic noise created by the structure of the English language—its vocabulary, its grammar, its idioms, its conventions.

Inexperienced writers in particular find it very difficult to go from an outline to polished prose with any degree of grace and ease. Most of them try to do too much in one step: choose the exact word; check for proper spelling and punctuation; test each sentence for balance, flow, and stylistic effect; and decide on all the mechanical matters of format. Since noise engenders noise, the cumulative effect is to produce *thesis paresis* or "writer's block."

We suggest that a practical substitute for experience is to divide the job of transcribing symbols into two operations: writing the draft, and revising. In essence this optimizes the functions of the encoder and the transmitter.

Writing the Draft

Writing a rough draft will enable you to transcribe notes into prose without having to worry about the refinements of language and usage, and the mechanics of composition and spelling. With it, you put your story together from beginning to end for that all-important first time. And you do it quickly, without stopping to doctor a sentence here and a word there, because you realize you are not producing a final version.

If your outline has entries that represent topic sentences of paragraphs, you simply have to "talk" your way through the outline. It will tell you what to say and the order in which to say it. You provide the continuity, or flow to your story as you unfold it.

Since the idea is to get through the draft as though you were giving a talk from notes, we suggest that you try dictation. Any dictating system that frees you from the physical act of writing will do.

Try to write without focusing your attention on the written words in front of you. (If you do both you are acting as transmitter and receiver simultaneously and may create undesirable feedback.) Concentrate on telling the story as though you were face to face with your intended reader. Say the words aloud if you wish. When you have to pause to think, don't look at what you have written unless you need a few words to help you start. If you read any more than that, you may stop to rewrite and this maneuver decreases the value of the draft.

When you finish, arrange to have a fresh copy typed. Ask the typist to leave wide margins and to triple space the text. You will then have a clean, uncluttered manuscript to work with.

Reviewing the Draft

The following notes on procedure should help you develop an effective technique for reviewing your draft.

1. Don't begin to revise the moment you finish writing. You have a much better chance of putting yourself in the reader's place if you get away from the writing for a while and come back fresh.
2. Review the reader's semantic needs and channel capacity. Any change you make in your manuscript must be justified on the grounds that it will help the reader. If he expects you to answer certain questions, be sure the answers are where he'll find them.
3. Divide the revising job into two parts: check the logic and flow of information first, then concentrate on polishing the writing. The first requires a continuous reading at a normal reading rate; the second is a stop-and-go operation.
4. As you read for logic and flow, mark all discrepancies but do not revise until you have finished the entire first reading. You also may mark any errors in composition and mechanics that catch your eye.
5. Reread each spot you marked in the first reading. Back up far enough to place the error in context, then determine whether you need to reword, delete, or add material. Correct accordingly.
6. Whenever errors are bunched, you may be able to revise easier by rewriting the entire sub-section or paragraph. Try to salvage the original first, but if you don't gain headway, talk your way through another version.
7. Save the mechanics of format until last. You will have detected most of the flaws, but a methodical check will take only a few seconds per page. Any omissions or inconsistencies can be corrected easily, even when you're tired. Several of the more common sources of mechanical noise were listed earlier. Here are some others:

    Captions and numbers not assigned to figures.
    Figures not referred to in text.

Figure references appearing too late to be of help.
Appendix material not referred to in text.

8. Finally, reread from beginning to end to see how everything fits together.

QUICK TESTS FOR PROPER ENCODING—EDITORIAL FUNCTION

An editor is a proverbial "man in the middle." His job is to assist both ends of the communication system—the transmitter and the receiver. In editorial conferences with the writer, the editor acts as a stand-in for the reader, advising how to eliminate pseudo choices that the reader might logically face.

Not all engineering departments have a staff of editors. If yours is one that doesn't, you still can benefit from the comments of anyone who you think has a channel capacity approximating that of your intended receiver.

Title and Abstract

You can test the coding of your main title and abstract by giving the abstract, without title, to a colleague unfamiliar with the project being reported. Ask him to read the abstract and then write a title of his own. Compare this title with yours. The wording does not have to be exactly the same, but the two should agree in scope and emphasis. If they do not, then either your abstract is faulty or your title does not accurately represent your report.

The Conclusions

You can test the coding of your conclusions by having someone with approximately the same background as your receiver's read your report up to the terminal section. Ask him not to read your conclusions, but to jot down the conclusions he feels are justified by the evidence presented. Then have him read your conclusions and comment on any disagreements.

The Evidence

Any technical description you may have had trouble writing also should be checked. Ask a friend to read the passage and then to recount it in his own words. If he has misinterpreted your meaning, go over the written version with him and correct whatever misled him. Descriptions of the operation or construction of a new device usually cause trouble.

SUMMARY

Every time you write a report, whether short or long, informal or formal, you involve yourself in many communication functions. You supply the information for the message, you encode the information into symbols, you transcribe the symbols into written signals, and you select the mechanical channel. Only the last two duties can you delegate to someone else; most of the time you are responsible for them all.

Once you realize that you can produce a successful communication only if you first learn to communicate with yourself, you are on the way to becoming a good report writer. You then will have a feel for handling the choices you meet at every turn and for applying the rules of writing that heretofore may not have made much sense to you.

Helpful Reference Book

Robert R. Rathbone and James B. Stone, *A Writer's Guide for Engineers and Scientists*. (Englewood Cliffs, N. J.: Prentice-Hall, Inc., 1962.) This book covers the major problems of report writing and offers exhibits that show how the problems are solved. It is written particularly for technical people.

chapter

6

ORAL REPORTING

Man's voice still is his most valuable communication transmitter. Daily it handles more person-to-person transmissions than all other systems combined; this is just as true in engineering as it is in man's broader social environment. Yet despite the important role we assign to the voice, we do less than we should to understand this unique system and to improve our efficiency in its use.

The purpose of this chapter is to examine oral reports in the light of what we already have discussed in previous chapters and to suggest ways to develop an effective approach to solving report problems.

COMPARISON WITH WRITTEN COMMUNICATION

Oral communication of technical subject matter has much in common with its written counterpart. Both are human systems, both draw their language from the same sources, both follow the same basic rules of logic and rhetoric, and both impose certain demands upon the originator:

1. He must have information to transmit.
2. He must have a legitimate purpose for transmitting it.
3. He must code his message to fit the channel capacity of his receiver (background, experience, vocabulary).
4. He must control the noise in the system.
5. He must coordinate his separate functions as originator: information source, encoder, and transmitter.

In addition to possessing these major similarities, speech and writing both fall into three general types: informal, semi-formal, and formal presentations. This order represents the frequency of use and the order of importance of the three types as far as communication is concerned. The importance of content is a variable factor, arbitrarily set by the originator or by the receiver.

There are, of course, many differences between the two systems, and it is with these differences that we now are mainly concerned. We should point out, however, that the basic principles of sound reporting, as discussed earlier, apply equally well to speaking as to writing.

Feedback

As mentioned in Chapter 4, a speaker can utilize feedback to adjust his message to meet the needs of his listeners. This is true particularly for informal, person-to-person reporting, such as discussions in group conference. But as the size of the audience (number of receivers) and the formality of the occasion increase, exchange between the destination and the source is confined more and more to post-mortem diagnosis—a short question and answer period after the talk. This delayed feedback is useful, but it does not help efficiency.

Unfortunately, feedback does have a minor drawback: it may lull the speaker into complacency. He may use feedback as a crutch and fail to do a good job with his original transmission ("Why worry? I'll have a second chance!"). This attitude is an example of psychological noise at work in the encoder and transmitter.

Coding

Speech in general imposes less stringent demands on its users than writing does. In informal and semi-formal situations, a speaker is face to face with his receiver and his message can be coded in a more personal style than he customarily uses in his technical reports. For example, the personal pronoun "I" is preferred to "the speaker"; word contractions appear at will; all sentences do not have to be complete grammatically; short words, short phrases, and short sentences are used regularly; and physical signals, such as gestures and facial expressions, supplement and often supplant spoken signals. Even in formal speeches, the encoder has the advantage of seeing his receiver and being seen by him, and this visual contact, although remote, can help to increase the channel capacities of both.

Transmission and Reception

Our voice is capable of producing all sorts of connotative signals by word intonation and inflection. To duplicate the same signals in writing we need to add qualifying words or to resort to mechanical devices, such as punctuation and typography.

Qualifying words:	"I am very disappointed that he couldn't come." (The voice could show disappointment through intonation, without the speaker saying, "disappointed.")
Punctuation:	"Who needs money!" (The exclamation point in place of the question mark qualifies the meaning.)
Typography:	"Turn the dial CLOCKWISE." (Underlining the word or putting it in italics would also indicate emphasis.)

Qualifying words in particular just cannot match the efficiency of the oral signals they replace. Many produce semantic noise and use up valuable channel capacity.

On the other hand, *uncontrolled* intonation and inflection of voice signals also can be responsible for a great deal of mechanical and psychological noise. The transmitter of written signals may misspell a word; but the voice transmitter may mispronounce it, utter it in a raucous tone, drawl it, or say it so softly that the receiver misses it.

Smooth flow of information is perhaps more critical in speech than in a piece of writing. In both systems, each word builds upon the preceding words, but it often is not possible for a listener to go back and "re-listen" if the thought seems illogical or incomplete. A reader, however, has some control over reading speed. He may not always be able to read at the speed he desires (because of poor encoding) but at least he can reread if he wishes.

Also worth considering is that once a speaker begins to talk, he is obliged to keep information flowing. He cannot make frequent erasures or stop to think things over as a writer can. Time is slipping by, his receiver is waiting, and he must make each word count.

A final significant difference is in the emotional state of the two receivers. A reader is not "captive": he can read all, some, or none of a report; he can read material in any order he wishes; and he can stop for a rest. A listener is captive—at least in body. He will do all he is expected to do as a receiver only when he is convinced that the speaker is doing everything within his power to make the "captivity" as pleasant as possible.

Channel Capacity

The longest formal speech engineers have to make usually runs about thirty minutes. This is the average time allotted for presenting a talk before a professional society. Suppose you had written a report on a design project that had taken many months and were asked to discuss your work. During the half-hour, you would not be able to cover all the details you brought out earlier in your written report, but would have to concentrate on certain main points, such as the operational features of your design or the construction details. The communication channel just would not have the capacity to carry all the information you might wish to transmit in the thirty minutes.

Experienced speakers avoid the danger of trying to say everything in one speech. They choose whatever main points they think will interest their receivers, establish an intent for their communication, and then expand only one, or possibly two, of these points. The rule of thumb to remember is that the chance of each receiver retaining any given detail in a speech decreases as the total number of details increases. The receiver's channel capacity is directly related to his memory capacity.

SUGGESTIONS FOR IMPROVING INPUT AND OUTPUT

Understanding the mechanical characteristics of the communication system one is using contributes to its effective use. In the analysis of human systems, however, communication theory must include many aspects of psychological and sociological theory. *People* write reports and *people* make speeches. Whether or not they communicate successfully with one another depends in part on how well they are able to adapt themselves to all the conventions and restrictions that their society, their profession, their training, their job, and their genes impose upon them.

We do not intend that the rest of this chapter serve as a handbook for public speaking. Good texts in this area can be found in abundance. We simply wish to present a brief discussion of the common human problems associated with oral communication and to suggest practical ways to solve them.

The Impromptu Speech

Let us look first at off-the-cuff reporting. Undoubtedly there will be occasions when you will be asked to speak without advance notice. Such situations can be difficult to meet gracefully if you have never worked out a general plan of attack. Within your tightly knit group of fellow workers, you seldom have to worry about something to say. Outside your group, you may find yourself in strange territory. The "natives" may not be as hostile to you as they were to Gulliver, but they will ask questions.

Naturally, you cannot anticipate all of the many questions you will be asked about your work nor predict the particular circumstances under which they will arise. But you can design a pattern of thought that will guide you to a sensible course of action.

A prime rule is that you must be a careful listener whenever you have the slightest inkling that you may be called upon to contribute your views as a design specialist or as an engineer. For instance, if you unexpectedly find yourself in conference with a managerial group, immediately establish the context in which your thoughts are to be received. Has a question been posed that will have to be answered? Has an hypothesis been stated that will require supporting evidence? Have facts been presented from which conclusions must be drawn? Are recommendations being sought for present or future action? These are some of the common lines of development that a discussion might follow. The opening remarks by whoever is running the conference should establish the line of attack.

Once you sense which direction the discussion will take, you automatically know whether your purpose should be to inform, to instruct, to evaluate, or to persuade. You then are ready to concentrate on what to say and may be able to anticipate specific questions. If you do not immediately recall specific points to bolster your answer, don't panic. You may think of something while you repeat the question and state the general answer. Redundancy helps to speed recall, since one idea often will open a succession of associations.

If other speakers precede you, jot down the important points they make —points that you may agree with, disagree with, or wish to amplify. Should one of them not be able to illustrate a point he has made, perhaps you can. Such comments always make a valuable contribution.

As a word of caution, do not attempt to think out in advance the entire coding of any statements you may wish to make. You are working under pressure to produce thought; the wording will come naturally (and you have a better chance to control noise) if you concentrate on the *key words* for the *key ideas*.

The Formal Speech

Delivering a speech before a large group is not an everyday occurrence for an engineer. But since it represents the most rigid situation you will encounter, we feel we should devote some time to it.

Before you can prepare a speech intelligently, you should be instructed in the following preliminaries. Any program chairman worth his salt should supply the answers before you have to ask for them. But do not rely on it. Most chairmen need prodding.

1. *Brief yourself on the receiver.* Who will make up the audience? How many will attend? What are their common interests? What do you have to say that will be of importance to them?

2. *Brief yourself on the physical surroundings.* What is the size of the meeting place? What facilities are available for visual aids? Is there a public address system? A lectern or stand for your notes? These items are all part of the mechanical channel.
3. *Brief yourself on the context in which your speech will be received.* Will there be a long line of previous speakers? If so, what will they cover? Will there be an opportunity for questions and answers? Will speakers follow you? What will their subjects be?
4. *Brief yourself on the time schedule.* How much time will be allotted to you? Approximately when will your speech come—in the morning, in the afternoon, or just before lunch? The time schedule of a speech often influences the amount of mechanical and psychological noise you have to contend with.

In preparing your speech, formulate a minimum plan that will involve memorizing only key statements. Rely on being prompted by notes and visual aids for the remainder. Do *not* write out your speech and then read it. Write it out if this will help your organization and timing, but stop there. Reading it will promote boredom and boredom will promote sleep. The reason for this chain reaction is that you create a channel mismatch. Technical writing does not lend itself smoothly to reading aloud to others, especially if the written speech is a technical report taken from the file and dusted off for the occasion.

Since it is outside the province of this book, we will not take the time to review the ABC's of delivering a speech. But here are a few additional pointers that might help you in preparing one.

1. Concentrate on producing a strong opening and a strong finish. If you lack experience, write rough drafts of both and read them aloud three or four times, polishing as you go. Then prepare notes from the revised script. Memorize a few key passages if you wish, but not the whole bit. By feeling sure of the opening and closing of your message you free yourself of one worry that creates psychological noise; by offering the receiver a strong opening and closing, you code your message to satisfy one of his channel characteristics (he requires more attention at the beginning and at the end of a speech).
2. Put the whole show through several dry runs. Practice with a representative audience whenever possible. Any noise you can reduce now costs you nothing, so listen to your receiver's comments and suggestions. They provide feedback in advance!
3. Spend additional time on your visual aids. Work out a commentary that flows smoothly and easily. Practice here also will give you assurance and thus reduce noise. The following section treats visual aids in greater detail.

VISUAL AIDS

Illustrations are important to any speech: verbal illustrations provide examples for the mind; visual illustrations appeal both to the eye and to the mind and thus greatly increase the receiver's channel capacity. You may not be able to use visual aids for every speech you make, but you should never pass them up simply because you cannot be troubled to prepare them.

Your choice of medium to illustrate your talk will depend upon a number of practical considerations. (Foremost among these is availability, but for logical reasons we will treat it as irrelevant here.)

1. The type of visual representation required (drawing or photograph, still or moving image, etc.)
2. The visibility factor (viewing distance of audience)
3. The cost of preparation and display
4. The time required to prepare

If you have an unlimited budget and a lot of time to prepare your aid, you also have a wide range of choices, including colored sound movies and closed-circuit television. But even if cost and time are critical, many simple, inexpensive devices now are available commercially that will greatly increase the variety and effectiveness of your presentation:

>Magnetic stick-on boards
>Felt stick-on boards
>Precut letters and symbols for the above
>Colored scotch tape
>Flip boards and felt-point pens

Also consider using working models, exhibits, and samples whenever the audience can see them easily. Physical evidence can be more impressive than pictorial representation but keep the items covered until time for display.

Since tables, curves, and lists are used with great frequency in practically every technical speech, they should be checked carefully for mechanical and semantic noise. The major cause of noise is poor visibility. Nothing to be read should be smaller than $\frac{1}{250}$ the distance from the farthest viewer. Translated into meaningful terms, for a reading distance of 50 feet, letters should be at least $2\frac{1}{2}$ inches high. Specifications for overall size of letters and symbols, length of line, spacing, and width of margins on charts are given in a pamphlet published by the American Standards Association: Pamphlet ASA Y15.1-1959. Helpful information on the preparation and projection of slides is given in Chapter 11 and in a booklet *Hints to Authors*, issued by the American Chemical Society (Bulletin 8).

SUMMARY

Although oral communication and written communication have many system characteristics in common, each has its own attributes that make it uniquely equipped to meet the requirements of different reporting situations. The major advantages of the oral system are its feedback channel; its rapid conversion of thoughts into symbols; its personal, familiar style; its controlled use of word intonation and inflection; its ability to increase channel capacity by putting words into a physical context; and its adaptability to visual illustration. A clear understanding of these characteristics and their proper use will help the originator produce an effective, efficient communication.

chapter 7

GRAPHICS IN ENGINEERING DESIGN

INTRODUCTION

In an earlier chapter, the fact was brought out that our visual sense is the most efficient information receiver of all of our senses. The hundreds of judgments and decisions that we must make to get through a single day of life are based more on what we see than on what we smell, hear, feel, or taste. The world of industry and science that is served by the engineer is highly oriented toward visual communication. In research, design, production, sales, services, and education, the predominant form of communication is through the written word and through drawings and photographs.

In the present sense *graphics* is a more appropriate word than is *drawing*. It is defined as "the art or science of drawing, especially according to mathematical rules as in perspective, projection, etc." The design engineer's interest in graphics should not be in the traditional, formal, detailed drawings (shop drawings) but rather in the informal uses which can provide a strong communication link, first with himself and then with all others who must contribute to or pass judgment on his efforts.

In this sense, graphics is a coding technique. Lines are drawn on a plane surface (the paper) in such a way as to represent an object or scheme. The person who made the drawing hopes that he or other people can gain useful information (recognition) from the drawing. The better the coding job, the easier is the recognition.

The rest of this chapter and Chapters VIII, IX, X, and XI are concerned with the communication aspects of graphics, the graphic forms most useful to the design engineer, and some techniques and tools of drawing—all intended to show how engineering graphics can best serve the design process.

GRAPHIC COMMUNICATIONS

Communication with Oneself

Often when a design engineer is confronted with the specifications for a new design job, he has an immediate mental picture of the problem and some possible solutions. If he is an extremely creative person, many solution images may flash through his mind. This is the time to take a blank piece of paper and start sketching. He should have the facility to record these images as quickly and as accurately as possible. Many will prove worthless upon further reflection. Others will bring out serious faults in reasoning—but, perhaps one will provide an exciting approach worthy of further study. A critic of modern womanhood has said that women talk constantly in order to find out what is on their minds. At the outset of a design project, the engineer should sketch constantly to find out what is on *his* mind. Projects seldom start with the writing of equations or the programming of a computer for definitive answers.

As ideas jell and the most promising ones come forward, numerical answers to important questions are needed—answers upon which to base decisions for future progress. Here again, familiarity with graphic methods of analysis can provide quick, order-of-magnitude numbers. Even the most precise instrument drawing cannot replace the analytical techniques needed for final answers to some design questions. But, there are many techniques of graphical mathematics and curve and chart drawing that will tell you if you are in the right ball park and can point out the areas which need closer analysis.

Communication with Others

Colleagues, detail draftsmen, supervisors, model makers, and patent lawyers are just a few of the many people who need information about a new project as the design process moves on. The audience and their personal interests are varied. Graphics plays an important part in the reception of a designer's work. He should know his audience and should be able to present his work in the most efficient manner, whether for the machinist who must understand and build a device or for a company executive who must decide whether to invest in the ideas or not.

GRAPHIC FORMS

What are the graphic forms most useful to communications in the engineering profession, and what is the theory underlying the use of these forms? Posing these questions in the language of information theory results in the further question, "What are the codes and the techniques of coding which will produce maximum transmission of information with minimum noise?"

Engineering drawing is a vast and highly detailed field of study. We are concerned only with the basic principles and with the techniques which can best aid the engineer throughout the entire design process. More detailed discussions can be found in the many available drawing texts, some of which are listed in an annotated bibliography at the end of this discussion.

The terms *engineering drawing* or *mechanical drawing* bring forth an image of carefully prepared, adequately dimensioned, thoroughly annotated drawings on translucent, vellum paper from which prints can be

Figure 1. The drawing forms of most common use to the engineer—*Orthographic* and *Pictorial*—and how a cube would appear in each form.

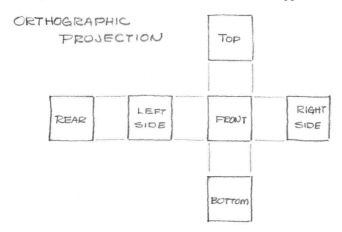

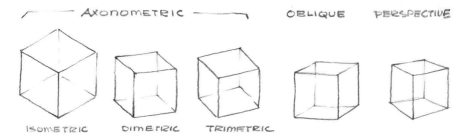

made for use in planning, purchasing, production, service, etc. Such drawings will not be discussed here. They are the realm of the detail draftsman and are of interest in the present context only in that the design engineer must be able to communicate his ideas to the draftsman as a prerequisite to the preparation of these drawings.

The graphic forms used by the engineering profession for depicting objects are *orthographic projection* and *pictorial drawing*. Figure 1 shows the types of drawing forms most commonly used and illustrates how a cube would look in each form.

The design engineer is not always concerned with objects or assemblies of objects. His activities often confront him with numbers, equations, relationships, and mathematical concepts. Charts, graphs, block diagrams, flow diagrams, and the techniques of graphical calculus and vector geometry are the graphical tools that can help him understand the data before him and communicate his findings to others.

An understanding of the theory and use of all of these forms is a prerequisite to using them effectively and confidently.

Orthographic Projection

Orthographic projection is an abstract representation of an object by multiple views. Traditionally, these views are the Top, Front, and Side (right-hand or left-hand) views. They are not realistic. Theoretically, they are the views one would see if observing the object from an infinite distance with the line of sight perpendicular to a given face of the object. Infinity is chosen as the vantage point so that rays projected from the observer to any point on the object are parallel. Figure 2 shows this configuration. In Figure 3, the object is enclosed in a transparent cube so

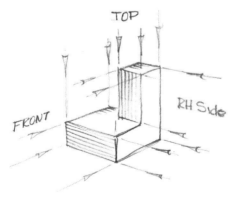

Figure 2. The basis for orthographic projection. Each (and any) facet of an object is viewed individually as if the observer were standing at some infinite vantage point.

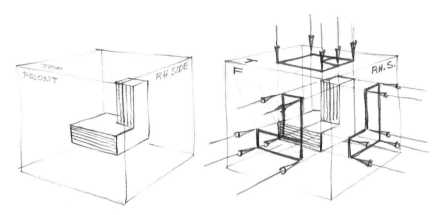

Figure 3. The frame of reference for orthographic projection is a transparent box enclosing the object. Orthogonal (right angled) rays from an infinite vantage point describe accurately the aspect of the object seen behind any face of the box.

that the faces of the cube are parallel to the principal faces of the object. If we now mark the points where the parallel rays pierce the cube, we obtain an exact image of each of the six faces of the object on the respective faces of the cube. When this enveloping cube is unfolded as in Figure 4 and laid out in a plane (the plane of the drawing paper), an orthographic projection of the object results. Only two views are required to define most objects. These two views are usually the Top and Front views. They describe the height (H), the width (W), and the depth (D) and show details on these two faces. The other four views, right side, left side, rear, and bottom, are redundancies as far as H, W, and D are concerned and are used only when needed to clarify details not shown adequately in the two principal views.

Many objects present faces that are not parallel to any of the six normal views. In these cases, *auxiliary views* are obtained by passing a projection plane parallel to the oblique face, projecting the details of the oblique face on the plane, and then unfolding it along with the principal views. Figure 5 shows the development of a single auxiliary view in conjunction with a top view and a front view. This extra view is a true view of the oblique face of the object. Reflection on this point will show that there are an infinite number of auxiliary views in the orthographic system. In other words, we may view an object from any one of an infinite number of vantage points about the object, and, by passing a plane perpendicular to our line of sight, record orthographically what we see from any one position. The selection of which view to show depends upon the configuration of the object.

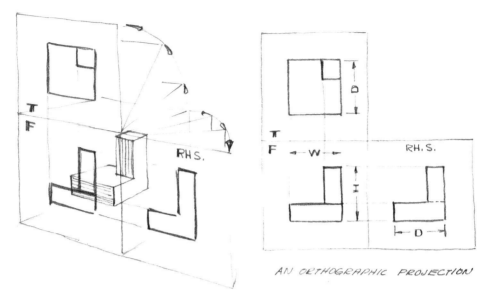

Figure 4. Unfolding the transparent box yields an orthographic projection—a two dimensional picture of the true aspects of the object showing its three principal dimensions: height (H), width (W), and depth (D). The other three faces (bottom, rear, and left-hand side) are not shown. The information they contain is redundant.

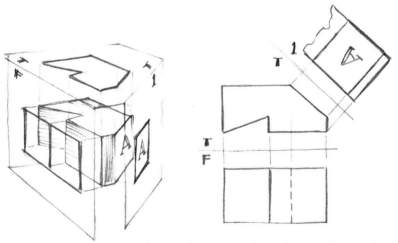

Figure 5. The development of an auxiliary view. View 1 was added to the two principal views (top and front) specifically to show a true shape of the oblique face A. Any number of auxiliary views can be projected as needed to record more information about the object.

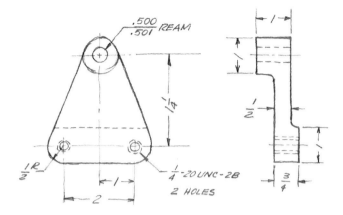

Figure 6. A dimensioned orthographic drawing.

Orthographic projection is used extensively in industry to depict parts for subsequent production. Each view is a true view (true scale) of one aspect of the part, i.e., top, front, right side, etc. Being true views, any details such as holes, angles, offsets, etc., can be dimensioned directly on the view. Figure 6 is a dimensioned, orthographic drawing. A skilled machinist could produce the part exactly as shown.

The design engineer has somewhat limited use of orthographic drawing. In most industries, he is not required to produce detailed drawings of the type shown in Figure 6. However, he must be familiar with the principles of orthographic projection in order to direct and check the efforts of a draftsman who is detailing his design.

In conceptual stages of a design project, the designer often resorts to simple orthographic freehand sketches to study the operating characteristics of a device or mechanism under study. Figure 7 shows a cutaway (or sectioned) view of a pneumatic valve. This type of sketch is usually done on rectilinearly ruled paper to aid in maintaining scale and proportion. If, as a result of his sketch studies, the designer decides that an idea looks promising, he may make a layout drawing using a drawing board and drawing instruments. Figure 8 shows such a layout drawing of a

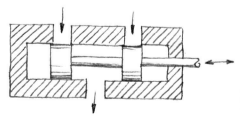

Figure 7. A freehand study sketch of a pneumatic valve.

77

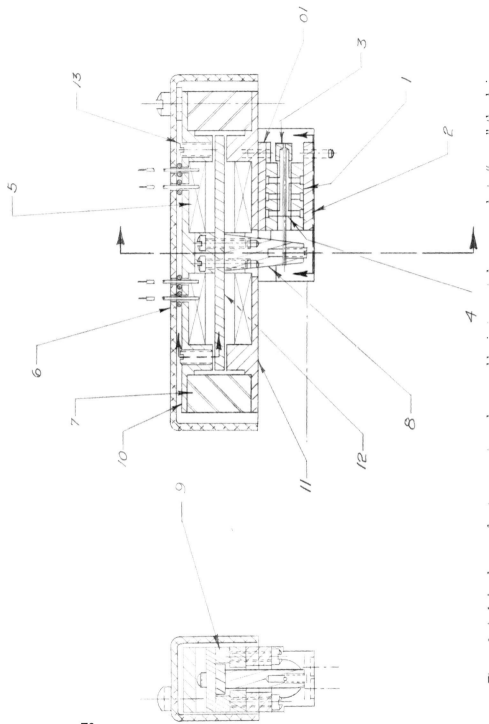

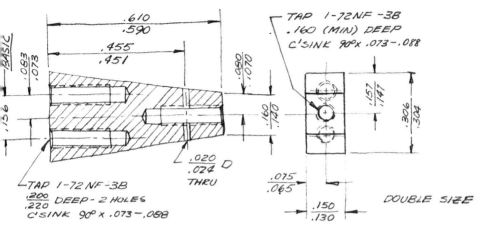

Figure 9. A freehand dimensioned sketch of a part to be made for experimental model verification.

torque motor and valve assembly. This drawing is made to establish dimensions and check operating characteristics. It is not a finished assembly drawing. There is enough information on this drawing and in the engineer's head to allow him to communicate with model makers, draftsmen, and others who can be of immediate help on the project. If he wishes to have an experimental model of his design built, he may quickly sketch a dimensioned orthographic drawing of each part of the device for the use of a machinist. Figure 9 shows such a detailed sketch.

Pictorial Drawing

As was mentioned in the previous section, orthographic projection is abstract—not true to life. Pictorial drawing on the other hand is an attempt to show solid objects as we actually see them. It is the next step below an actual photograph.

Figure 1 showed the various forms of pictorial drawing useful to the engineer. Of these, the axonometric and oblique forms are approximations to reality and contain distortion. They are easy to draw and are true scale. Perspective drawing is photographic imitation. It is somewhat more difficult to draw and does not result in a scale drawing. However, our purposes here are to show drawing as an efficient communication tool. Perspective, since it *does* correctly imitate what we actually see, will be stressed as a powerful device for the design engineer.

Before proceeding to the theory of pictorial forms, we must understand the idea of *foreshortening* and its effect on drawing. Suppose you are

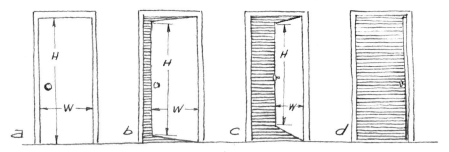

Figure 10. The effect of foreshortening. The door appears to change size and shape as it swings away from vantage point.

standing directly in front of a closed door. You would see the door as shown in Figure 10a. If asked to name the height and width of the door, you could make a good estimate based upon what you see. Suppose now that someone starts to open the door from the other side. You soon see the door as shown in Figure 10b. The height and width (and even the shape) of the door no longer appear as they were. The top, bottom, and left edges now appear shorter than when the door was closed. The top and bottom lines are foreshortened because they are no longer perpendicular to your line of sight. The vertical left edge appears shorter because it is now farther away from you. In Figure 10c, the door has been opened still further and these lines are shorter still until the door stands fully open, at Figure 10d, where the dimensions become minimum (from your vantage point).

Common sense tells you that the door has not changed size or shape. Foreshortening is that illusion by which a line (or dimension) on an object changes size depending upon our vantage point with respect to the object. We see a dimension in its true size only when the dimension is perpendicular to our line of sight. (Strictly speaking, this statement is not true since objects become smaller the farther away we are from them. For our purposes, however, we will consider scale as a relative

Figure 11. Common examples of foreshortening.

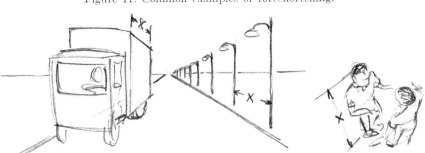

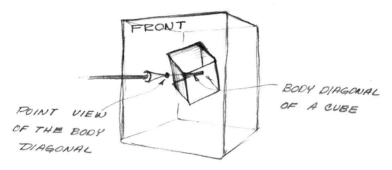

Figure 12. Space condition for the creation of the isometric pictorial form.

quantity. If all parts of a drawing are to the same relative scale, then the magnitude of the scale is unimportant.) Figure 11 shows three common examples of foreshortening. The dimension marked "X" in all cases is seen as being much shorter than we know it actually to be.

In Figure 1, it was shown that *axonometric drawing* has three forms, isometric, dimetric, and trimetric. Isometric drawing is the most common pictorial form used in engineering. It is the easiest to draw. All three principal axes (those that show height, width, and depth) are foreshortened equally. In dimetric drawing, two of these axes are foreshortened equally and in trimetric drawing, none of the axes have the same foreshortening.

The axonometric forms are derived from a special case of orthographic projection. If a cube is positioned within the framework of the six principal orthographic planes such that the body diagonal is perpendicular to the frontal plane, the resultant front view is an *isometric projection*. Figure 12 shows this arrangement. The same effect can be obtained by looking at the front view of a cube (Figure 13a), then rotating the cube about a vertical axis 45° (Figure 13b), and then tilting the cube forward 35°16′ (Figure 13c). The body diagonal, *AB*, now appears as a point, or is perpendicular to the frontal plane.

Figure 13. Orienting a cube to show a point view of a body diagonal—result: an *isometric projection*.

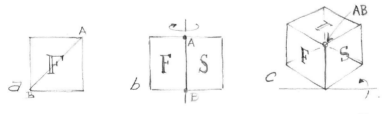

81

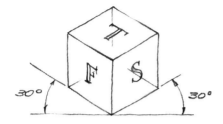

Figure 14. An *isometric drawing* of a cube. Foreshortening has been neglected resulting in a much larger appearing cube. (Compare with 13c.)

Examination of this figure shows that the three axes representing H, W, and D form a 120° angle with each other and that these lines are foreshortened equally (being $\frac{81}{100}$ of true size). These facts allow us to set up a form for *isometric drawing* as shown in Figure 14. The three axes are easily drawn with instruments (a T-square and a 30-60-90 triangle). If we neglect foreshortening, we eliminate the need for calculating $\frac{81}{100}$ of each dimension we wish to draw. The only effect this has on the resultant drawing is that the isometric view looks larger than the object actually is (compare the visual sizes of the cubes in Figures 13 and 14).

Dimetric and trimetric projections can be obtained in the same manner as was done for the isometric form. In dimetric, the cube is oriented so that two axes have the same foreshortening. A standard form for *dimetric drawing* has developed as shown in Figure 15. The axes foreshortening ratio are $1:1:\frac{1}{2}$ when the angles of the two receding axes with the horizontal are as shown. This form has an advantage over isometric drawing in that one side is displayed more than the other—a distinct advantage when depicting long, relatively thin objects. It is harder to draw since the angles do not agree with standard drawing instruments.

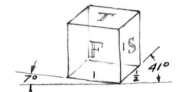

Figure 15. A dimetric drawing of a cube. The axes are scaled to a ratio of $1:1:\frac{1}{2}$.

The *trimetric drawing* form will be neglected here in favor of perspective drawing since, with three separate scales to contend with and odd-angles for the axes, it is difficult to use.

Oblique drawing is derived from a projection system in which the projectors do not make 90° with the picture plane. The result is a pictorial in which one face of the cube is parallel to the picture plane and is thus a true view. Figure 16 shows two forms of oblique drawing, a *cavalier* drawing at (a) and a *cabinet* drawing at (b). If most of the detail on an object is in one plane (or a set of parallel planes), the oblique system has an advantage in that the detail plane can be oriented parallel to the true-view face of the system. The object in Figure 16(c) shows this. The face with all the holes was placed in the true view plane and thus the holes all appear as true circles. The cavalier drawing shows much distortion. In the cabinet drawing, the distortion is reduced but it is necessary to reduce scale on the receding axis.

Perspective is the pictorial form of drawing that shows an object exactly as we see it. Figure 1 shows a perspective view of a cube. Note that the edges of the cube, which we know to be parallel, are not parallel in the drawing but are converging slightly toward the back of the cube. As a result, the front vertical edge, the line closest to the observer, is the only true length line on the cube. All others are shorter than true length

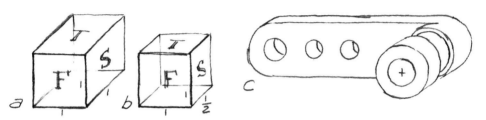

Figure 16. Oblique drawing. (a) cavalier, axes 1:1:1; (b) cabinet, axes 1:1:½; (c) the principal advantage of the oblique form—parallel circles can be drawn as circles rather than ellipses.

because they are either foreshortened or are farther away. The result, however, gives the most realistic view of the cube of all the pictorial forms presented.

The theory of perspective projection is complicated and beyond the scope of this discussion. We will, however, look at the basic principles of perspective and then, in the next chapter, show how these can be used to make realistic freehand sketches.

You would be making a perspective drawing if you stood in front of a window and, without moving your head, drew on the glass the scene you saw outside the window. This is how perspective is formed.

An observer stands in front of an object, places a picture plane (windowpane) vertically between himself and the object and draws what he sees on the plane. Figure 17 illustrates these conditions. We relate an object in our line of

Figure 17. Tracing an outdoor scene on a window pane would be creating a perspective drawing.

83

vision to a horizon line. This is the horizontal line that appears in front of us when we gaze directly ahead. It is actually the intersection of a horizontal plane extending from our eye level (horizon plane) and a vertical picture plane as shown in Figure 18. The horizon line extends to infinity to the right and to the left. Figure 19 shows a one-point perspective. Horizontal lines that are parallel on an object (lines ab and xy in Figure 19) converge to a point on the horizon line. This point is called the *vanishing point*.

Figure 20 shows a projected two-point perspective of a simple object. Note that the two sets of parallel horizontal lines converge to form two vanishing points (VP_L and VP_R) on the horizon. Technically, the third set of parallel lines, the verticals, should converge to a third vanishing point below the ground forming a three-point perspective. Since most objects drawn by engineers are relatively small, ranging in size from a tiny set screw to an object the size of an automobile, the third vanishing point can be neglected and the vertical lines kept vertical. Horizontal

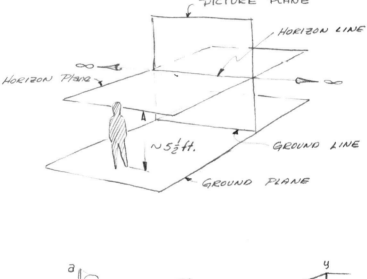

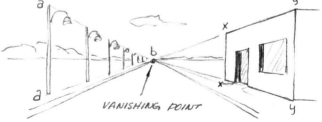

Figure 19. A one-point perspective sketch. Horizontal parallel lines converge to a *vanishing point*.

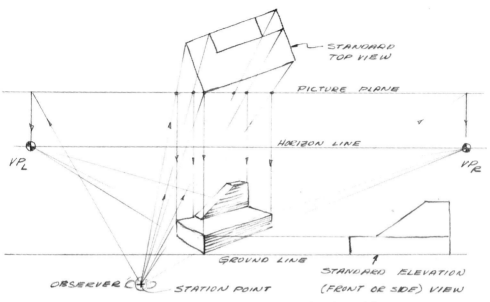

Figure 20. A projected, two-point perspective drawing derived from top and front orthographic view of a simple object.

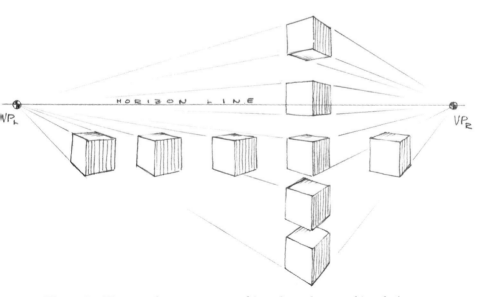

Figure 21. How an observer sees an object depends upon his relationship with the object. Here are nine different aspects of a cube, sketched using the same vanishing points.

lines on the object that are not parallel to the lines defining the object outline recede to the horizon at a new vanishing point. Oblique lines on the object have vanishing points some place other than on the horizon. The construction required to find these vanishing points is tedious. These lines can be easily found by determining the two ends points and then joining these points with a straight line. Many of these fine points of perspective projection are described thoroughly in drawing texts. The intent here is to show fundamental ideas and then, in the next chapter, show how these ideas can be used to make effective perspective sketches.

Naturally, a different view of an object can be obtained as the observer changes his position with respect to the object. Figure 21 shows some of the views of an object as the observer moves relative to the object (or as the object is moved relative to the observer).

chapter
8

HOW TO SKETCH

Drawing is a skill, and, as with any skill, proficiency comes with practice. The previous chapter presented the theory of the drawing forms most useful to the design engineer. This chapter will show how these forms can be used throughout the design process. Freehand sketching techniques are emphasized. Graphics' main contribution to the design engineer is as a means of communicating with himself—finding out what is on his mind. Except for the drawing of a careful design layout or an analysis by precise mathematical construction, instrument drawing is a waste of time for the engineer *if* he is well versed in the techniques of quick and accurate freehand sketching. These techniques are not difficult to master and one need not become an art student to learn to sketch effectively. All that is needed is an understanding of the form and *practice*.

TOOLS OF SKETCHING

The tools required for freehand sketching are, simply, paper and pencil. Engineers are notorious for making idea sketches on the back of purchase requisition forms, paper napkins, table tops, damask table cloths, etc. Admittedly, these media are often difficult to file. The point is that when an idea strikes a truly creative person, he will look for the nearest thing at hand on which to record or try out his idea. Any paper will do for sketching. A large sheet (17 in. by 22 in.) is handy because it does not limit the scope of your drawing and still folds down to file size. Translucent vellum is not the best surface for sketching, but will allow copies to be made by the diazo process.

The HB grade drawing pencil is ideal for sketching. It is a soft pencil so that lines can be made to vary from very light guide lines to dense black finished lines just by varying the pressure. The ordinary office-types, No. 2 or $2\frac{1}{2}$, are also good for sketching. Drawings made with these "lead" pencils will smear if rubbed. Some engineers prefer a wax base pencil such as the Eagle Verithin or the Eagle Prismacolor. These pencils yield very light greys up to dense blacks and the drawings will not smear. Erasures are difficult to make when using a wax pencil. This same disadvantage applies to sketching with fountain pen or ball point pen, but some people prefer the freedom and type of line that a pen gives. The best solution to the problem of tool selection is to experiment with various papers and pencils or pens and then choose the ones that suit your needs.

SKETCHING

Before studying the techniques of sketching in the orthographic, axonometric, oblique, and perspective forms it is well to look at some qualities that are common to all forms. These qualities are the following: scale and proportion; blocking-in of objects; visualization and drawing-through; and the handling of circles, circular arcs, and curves.

Scale and Proportion

For most freehand sketching needs, scale is unimportant. Scale is a relative quantity. If all parts of the drawing are to the same scale, the information contained in the drawing is readily received whether the drawing is larger than, smaller than, or equal to actual size. *This fact makes proportion the important consideration in sketching.* If you wish to show a box 3 in. high by 5 in. wide by 2 in. deep, the scale used is unimportant but you must maintain the proportions of 3:5:2 for the height, width, and depth in order to transmit the desired information successfully.

Proportion and scale are easy to maintain in orthographic and isometric sketching, especially if printed grid papers are used. In perspective sketching, however, foreshortening must be taken into account to maintain visual proportions. Problems of proportion will be discussed with the respective sketching forms in succeeding sections.

Blocking-In

Any object may be enclosed in a rectangular prism that describes the basic dimensions of height, width, and depth. Figure 22 illustrates this point. In sketching, it is most advantageous to sketch this prism first, in the correct form and to correct proportions, and then to build details

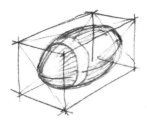

Figure 22. Any object can be enclosed within one or more rectangular prisms. These prisms establish height, width, and depth for the finished drawing.

within the prism. For extremely irregular shaped objects, more than one prism might be needed to effectively contain the object.

The designer interested in improving his facility at sketching should make a habit of carefully constructing this outline prism first. It will enable him to maintain correct proportions easily and will serve to keep the entire sketch in the desired drawing form. Too many people attempt to sketch an object by starting at one corner and developing details as they go along. Before they are half-way through, they have lost the scale, proportion, and form of the drawing.

Visualization

Visualization is the ability to see through and around an object—to know, or imagine, what is on the hidden sides. A pictorial view of an object normally exposes three of the six principal sides. The design engineer should understand what is on the hidden sides and the relation of these sides to the exposed sides. The technique of blocking-in aids in this understanding. If, in drawing the basic prism, the hidden lines are put in lightly (not as dashed lines in the conventional sense), the designer will be constantly aware of the three-dimensional nature of the object. Figure 23 illustrates this point.

Children and primitive painters do not have this ability to visualize in three dimensions. When asked to draw a house and a cow, they produce

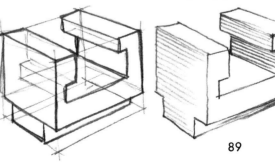

Figure 23. *Drawing-through* helps establish proportion and shape in a pictorial sketch.

Figure 24. Young children and "primitive" painters draw three-dimensional objects in a flat two-dimensional manner.

drawings as shown in Figure 24. They know instinctively that a house has four sides and that a cow has four legs, but they have not gained the ability to visualize in the third dimension. Their drawings, therefore, are two-dimensional, but with a serious attempt at reality.

Circles, Arcs, and Curves

All objects are made up of a combination of straight lines, circles, circular arcs, and irregular curves. In instrument drawing, curved lines are handled readily with the compass and the french curve. In freehand sketching, however, means must be found by which curved lines can be quickly and accurately placed. The technique for doing this is the same for all drawing forms. Inscribe a circle in a square, as shown in Figure 25, and draw center lines perpendicular to the sides of the square. The circle is tangent to the square at points A, B, C, and D. This knowledge is all that is needed to sketch circles in any form, since the square and the center lines can be duplicated easily in the orthographic or pictorial

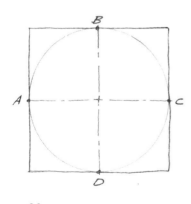

Figure 25. Circle construction basic to all pictorial drawing forms.

forms. Figure 26 shows the procedure for sketching circles in orthographic, isometric, and perspective drawing forms.

A circle viewed obliquely yields an ellipse. Figure 27 shows the changes in a circle from its true view, through various ellipses to the straight line or edge view. Many objects in engineering design work are cylindrical. A pictorial drawing of a cylindrical part is made up of a number of ellipses joined by straight lines tangent to the extreme edges of the

Figure 26. Sketching a circle in (a) orthographic, (b) isometric, and (c) perspective forms. Note that in all forms, the center of the enclosing "square" can be found by sketching the diagonals.

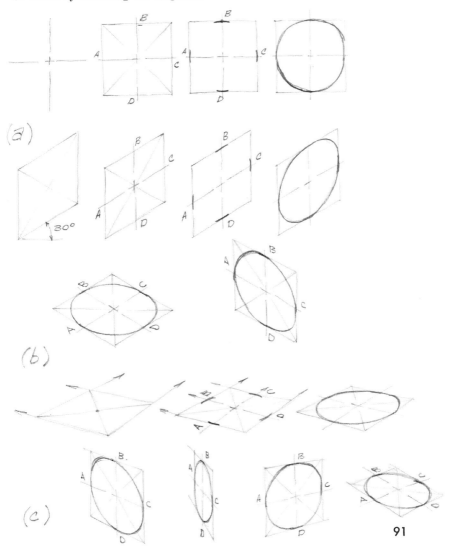

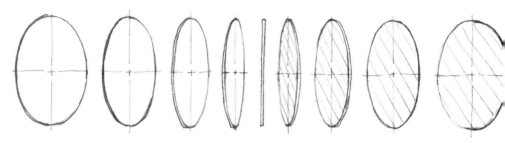

Figure 27. Various aspects of a circular disc from a straight line (edge) view, through various ellipse forms approaching the true (circular) view.

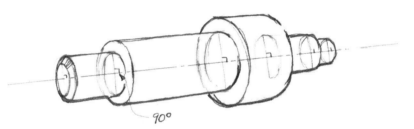

Figure 28. A perspective sketch of a cylindrical part. Note how "drawing through" helps control the sketch and also that the major axis of each ellipse is at right angles to the axis of the part.

Figure 29. A common error in perspective sketching. At (*a*) the major axes of the ellipse are vertical. At (*b*) they are correctly positioned perpendicular to the axis of the part.

ellipses as shown in Figure 28. Note that the major diameter of each ellipse is perpendicular to the axis of the circle represented by the ellipse. *This is an important rule in sketching and will do more toward creating a successful and satisfying sketch than any other rule.* A common mistake is to draw ellipses as shown in Figure 29a where the major axes of the ellipses are vertical. Figure 29b is a correction of this fault according to the above rule and results in a much more satisfactory and realistic sketch.

After considerable practice at drawing pictorial circles (ellipses) by first blocking-in the circumscribed square and finding the four tangent points, you should be able to sketch an ellipse in place with no construction. The "fatness" of the ellipse (ratio of major diameter to minor diameter) can be estimated by eye and the correct position can be determined by making the major diameter perpendicular to the axis through the center of the circle.

Circular arcs such as are found on rounded ends of objects are handled in the same manner as full circles. The entire enclosing square (or parallelogram) need not be drawn. Figure 30 shows an irregular object made up of straight lines and circular arcs. Also shown is a perspective of this object showing the construction used to obtain the arcs.

Irregular curves can be sketched by using coordinate methods or, quite often, by eye. Figure 31a shows the orthographic views of an object with an irregular curve. Figure 31b shows an isometric sketch of this object.

Figure 30. In pictorial sketching, portions of circular arcs are developed in the same manner as the complete circle.

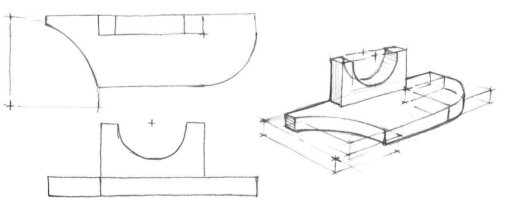

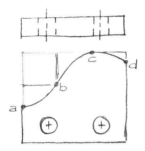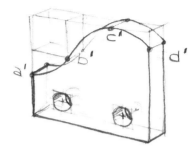

Figure 31. Irregular circles can be defined as a series of coordinate points on an orthographic view. These coordinates can then be transferred readily to a pictorial sketch, yielding an accurate pictorial curve.

Coordinate points a, b, c, and d were established on the orthographic view and then transferred to the isometric in the same manner as any other dimension. This resulted in the points a′, b′, c′, and d′ through which a smooth curve was sketched. Points on the front curve were then projected back along the right-receding axis of the isometric a distance equal to the thickness of the part giving four points on the back curve.

Orthographic Sketching

The procedure for making an orthographic sketch is shown in Figure 32. The object to be sketched is shown pictorially. The steps are as follows:

1. Block in the top, front, and right side views according to orthographic projection and the basic dimensions of H, W, and D.
2. Draw center lines for all circular parts.
3. Lightly sketch in circles and circular arcs.
4. Establish other details of the object.
5. Finish off the object with dark, clean lines.

Isometric Sketching

Figure 33 shows an object to be sketched in isometric form. The procedure is as follows:

1. Establish the three isometric axes.
2. Lay in the enclosing prism by making (or estimating) full-scale measurements of H, W, and D along their respective axes. Sketch in the enclosing prism by drawing lines parallel to the principal axes.

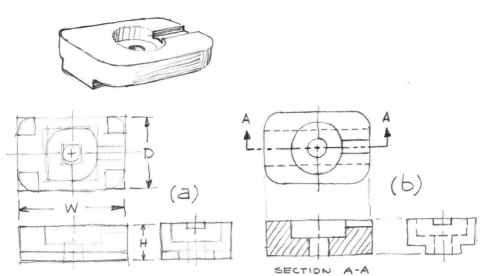

Figure 32. Steps in making an orthographic sketch. (a) Blocking in of H, W, and D, finding center lines and establishing details. (b) Finished sketch—the front view has been sectioned (front half removed) to give clarity to interior details.

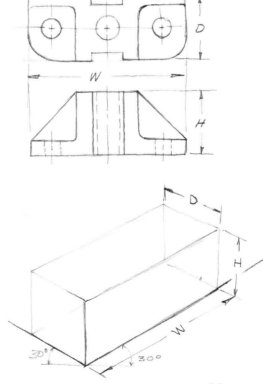

Figure 33. Steps in making an isometric sketch. (a) Block in H, W, and D on the isometric axes.

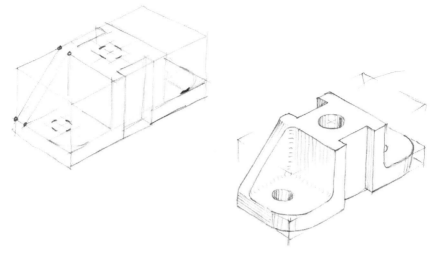

Figure 33. (b) Add details within the enclosing prism using full-scale measurements on all dimensions parallel to the isometric axes. Find non-isometric lines by coordinate methods. (c) Finished sketch.

Note that the hidden lines of the prism are included to help give a feel for the solid nature of the object.
3. Lay in the enclosing parallelograms that define the circular parts of the object. Note the points of tangency with short, tangent lines.
4. Sketch the details of the object lightly. Lines *mn* and *op* are non-isometric lines. They are found by first finding points *m* and *n* and *o* and *p* through direct measurement along principal axes. The straight lines *mn* and *op* can then be drawn.
5. Finish the drawing with dark, sharp lines.

Procedure for making dimetric and trimetric sketches is exactly the same as for isometric. The axes are positioned differently and the scale ratios for measurements along the principal axes are different.

Oblique Sketching

As was pointed out earlier, the oblique drawing form, and especially the cabinet form, is a special case in which one face of the object is parallel to the picture plane resulting in true views of any face parallel to the picture plane. This is very useful in drawing objects that have many parallel circular elements since, if these elements are oriented parallel to the picture plane, the circles will appear as true circles rather than ellipses. Consider the part shown orthographically in Figure 34. It is desired to make a cabinet drawing of this part. The procedure is as follows:

1. Establish the cabinet axes according to rule.
2. Lay in the enclosing prism by measuring H, W, and D along their respective axes. Remember that the receding axis is ½ scale in cabinet drawing.
3. Determine the axis of the object and find centers along this axis for the circular elements. (Observe ½ scale for these measurements.)
4. Draw the circles lightly to the correct radii.
5. Establish outlines of object by straight line connectors tangent to the respective circles.
6. Darken all visible lines.

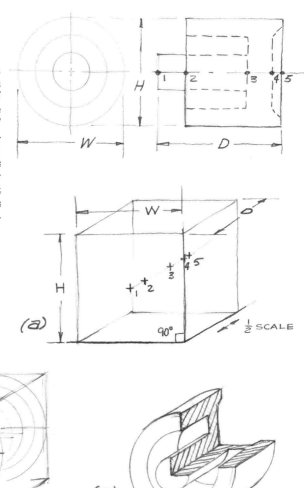

Figure 34. Steps in making an oblique sketch. (a) Establish the three oblique axes; lay off H, W, and D (½ scale on the D axes indicates the "cabinet" form of oblique); and establish the five arc centers. (b) Draw all the circular arcs lightly. (c) Finish drawing by connecting arcs with tangent lines. An upper quadrant has been removed to clarify interior detail.

Should an object contain circles in the top and side planes as well as the front, the procedures used for isometric circles can be used. The result, of course, will be ellipses.

Perspective Sketching

This chapter, so far, has presented freehand sketching (and, in particular, pictorial sketching) as the most powerful tool of engineering graphics. An understanding of a few basic principles and a lot of practice can give anyone the ability and confidence to make realistic sketches of simple, geometrical objects quickly and accurately. Perspective, since it offers a realistic, or photographic, representation, is the most suitable form of drawing for the clear transmission of engineering information.

A student or an engineer interested in improving his skill in graphical communication would do well to concentrate on the perspective form. Actually, as will be demonstrated here, perspective sketching involves only minor deviations from the more geometrical forms of isometric and oblique. In freehand sketching, since we do not rely on drawing instruments as aids in making precise lines, correct angles, or perfect curves, any advantage gained by simplicity of form is minimized. We are substituting the simplicity of one form for the realism of another.

The steps in sketching an object in the isometric and oblique forms are (1) establish the three principal axes, (2) block in the enclosing prism with scale measurements along the three axes, and (3) add detail and finish. For perspective sketching, we modify the isometric or the oblique axes according to the particular view that we wish to create and we consider true scale on only one axis. To complete the blocking-in process, we must understand foreshortening and have the ability to proportion by eye rather than estimate scale dimensions.

Consider the object in Figure 35, shown in orthographic projection at (a) and in isometric at (b). To sketch a perspective of this object, we must first choose a point of view from which to observe it. Most of the detail is on the front and the top so we should look down on the object and orient it so that we see more of the front than we do of either one of the sides. This information allows us to choose the three principal axes OA, OB, and OC. Note that the angles α and β are not equal as they are on the isometric and that axes OB and OC lead off to vanishing points VP_R and VP_L respectively (not on the page). We now lay off the height, Oa, of the enclosing prism on axis OA. *This is the one, and the only one, true scale line on the drawing.* Oa need not be true length but, whatever scale is used, its length establishes the scale of the drawing.

We now find points b and c and erect verticals. Distances Ob and Oc are proportioned from true scale taking into account the rules of foreshortening. Distances Ob' and Oc' are the true scale lengths for the width and depth of the object respectively. Note that Oc is foreshortened more than Ob. The amount of shortening of the two receding axes is in direct proportion to the size of angles α and β. (Remember the opening door in Figure 10.)

We have enough information now to complete the enclosing prism. Lines on the object that are parallel to Ob converge toward VP_R and lines parallel to Oc converge toward VP_L *even though we do not know where these two vanishing points are.* This is an important step in perspective sketching. These lines cannot be diverging, parallel, or greatly converging. If they converge too fast (if the VP's are close together), considerable distortion will result.

We may now add the details of the object always keeping in mind the foreshortening of dimensions along the receding axes and the fact that parallel horizontal lines on the object converge to vanishing points to the right or the left (see Figure 35d). Lines such as mn and op vanish at a point in the air. It is not necessary to find this point, however. As with isometric sketching, it is only necessary to find points m and n and points o and p and then join them with straight lines. If the construction is

Figure 35. Comparison of the orthographic, isometric, and perspective forms of drawing.

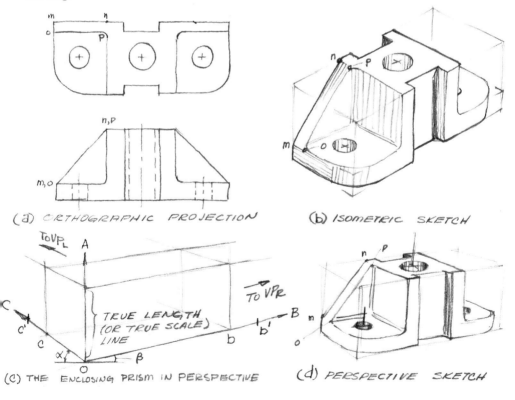

reasonably accurate, these two lines will converge to a point above the horizon and off the page.

Circles and arcs are developed by (1) blocking in the enclosing square, (2) finding the equivalent perpendicular bisectors of the sides and (3) observing the four tangent points thus obtained while sketching the ellipse. Note that the axes, through the center of the circles and perpendicular to the plane of the circles make 90° with the major diameter of the ellipse. This agrees with the rule presented in Figures 28 and 29. Accurate construction will automatically assure that this rule is not violated. Conversely, it can be used as a check on the construction to determine if circles are correctly laid into the perspective form.

Selecting a Point-of-View

A further word should be said about selecting a point-of-view for a perspective sketch. This selection is related closely with the size of the object and its relation with the ground plane. Figure 36 shows five objects, a bottle, a card table, a standing human being, an automobile, and

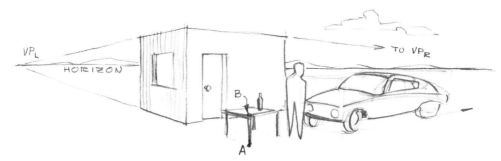

Figure 36. Using the human figure to establish scale in a perspective sketch. Here the observer is standing directly in front of the scene. Any point that lies on the ground (such as point A) is $5\frac{1}{2}$ feet below the horizon.

a small building, sketched from a position about 20 feet away. The observer is standing erect on the ground looking directly at the scene. In the section on the theory of perspective (Chapter 7), it was shown that the horizon line is a horizontal line directly in front of our eyes as we gaze straight ahead. The distance from the ground to eye level of the average human being is about $5\frac{1}{2}$ feet. This fact was used to establish the scale of Figure 36.

Assuming that the person in the scene is of average height, his eye level is $5\frac{1}{2}$ feet above ground. Therefore, since he is standing on the

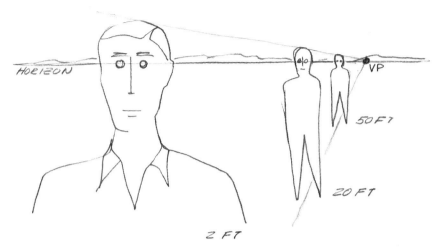

Figure 37. In observing another average-sized erect human, note that his eye level is always on the horizon no matter how far away he is.

ground, he must be positioned so that his eyes are on the horizon line. His height is determined by the distance between him and the observer (20 feet). We have a cone of vision, when gazing in one direction, of 30°. If the man in the picture were 2 feet away from us, we would see only his head and shoulders. At 20 feet, we see the entire figure and at 50 feet we also see the entire figure, but smaller. Figure 37 illustrates this. Note, however, that in all three situations, the man's eye level is at the horizon line.

Returning to Figure 36, we have established the ground level by estimating the height of a human being as viewed from a 20-foot distance, and we can now make good approximations of the heights of the other objects. The card table is 30 inches high, the bottle is 10 inches high, a modern automobile is about 5 feet high, and the one-story building is about 10 feet high. The human was included to emphasize the point that the horizon line is always $5\frac{1}{4}$ feet from the ground at all points in the picture.

Always start a perspective sketch by drawing a horizontal line, the horizon. Then, no matter where we start the perspective of an object located on the ground, we know immediately that the vertical distance from the base of the object to the horizon line is $5\frac{1}{2}$ feet. Thus, if we start to draw the 30-inch high card table by selecting point A, we know that point B must be about half way between A and the horizon.

The foregoing discussion assumes that the observer is standing on level ground. Suppose we raise our point of view 10 feet as we would if we viewed the scene from a second-story window. We would view the scene as shown in Figure 38a. Our horizon line is still at our eye level, which is now 15½ feet above ground. Our human friend's eye level is still 5½ feet above ground. Conversely, if we view the scene from the bottom of a 10-foot hill, we would see it as shown in Figure 38b.

We now have a means of estimating vertical distances on a perspective sketch. As was pointed out earlier, we neglect the foreshortening of vertical lines for the purposes of engineering sketches. Look now at the problem of selecting a horizontal position with respect to the object in order to obtain the most desirable view.

Figure 39 shows six cubes, each cube presenting a different aspect of the visible surfaces. The same vanishing points were used for all six cubes. The distance between the vanishing points has been divided into four equal segments. *It can be proven that the best locations for an object being presented in perspective form lie between the two vertical lines drawn through points M and N and at distances from the horizon (above or below) such that $\theta > 90°$*. Positions which vary greatly from these two conditions will result in a distorted perspective.

Figure 38. The effect on the perspective sketch as the observer changes his height with respect to the observed scene. (*a*) The observer stands 10 feet above the scene and (*b*) 10 feet below.

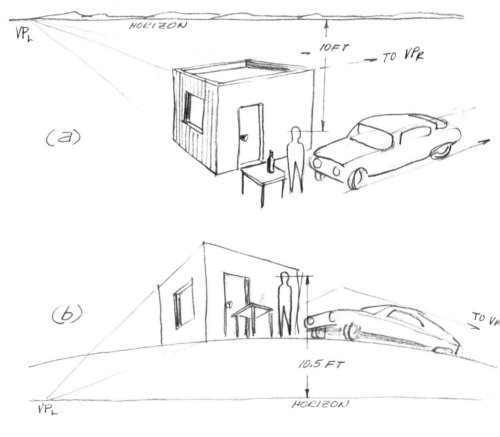

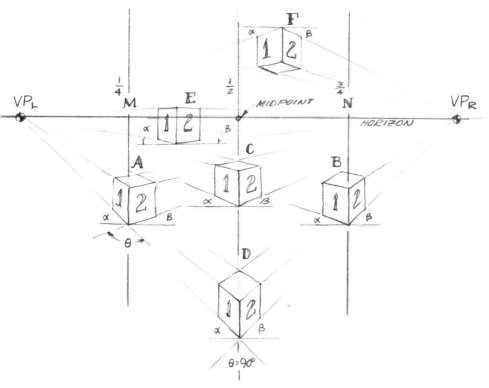

Figure 39. Criteria for establishing distortion-free perspective sketches: 1. keep the object in the area between vertical lines M and N (the middle half of the distance between VP's) and 2. set α and β such that $\theta \geqslant 90°$.

Note, in Figure 39, the extent to which the surfaces of the cubes are exposed as a function of the angles α, β, and θ. On cube A, side 2 predominates and $\alpha > \beta$. Cube B is the opposite, with side 1 more exposed to view and $\alpha < \beta$. Cubes A and B are on the two vertical lines through points M and N. Moving cubes A and B outside of these lines would introduce distortions.

On cubes C and D, sides 1 and 2 are exposed equally and in both cases $\alpha = \beta$. The difference is that cube D shows more of the top surface than does cube C. Cube D is farther away from the horizon, and, since the horizon represents our eye level, we are looking down on cube D more than we are on cube C. The angle θ at D is 90°. Moving cube D farther away from the horizon would make $\theta < 90°$ and would distort the view. Cube E straddles the horizon and represents a relatively small cube held at eye level or a distant cube that is taller than the observer. Cube F demonstrates the case of seeing two sides and the bottom of an object held above eye level.

An understanding of the role that angles α and β play in setting up a perspective sketch allows us to proceed confidently with the sketch without actually placing and using the two vanishing points. For an object drawn at a reasonable size on standard $8\frac{1}{2}$ by 11-inch paper, the vanishing points should be at least 24 inches apart. Obviously, one must have a larger sheet in order to have the VP's in the drawing or a means of estimating the position of the VP's. Angles α and β provide means for doing the latter.

Figure 40. Making a perspective sketch. (a) The basic form and dimensions are set up. Angles α and β are set depending upon how the object is to be viewed. Dashed lines are not drawn, but merely estimated. VP's are imaginary. (b) From the construction of (a) the details of the object are added.

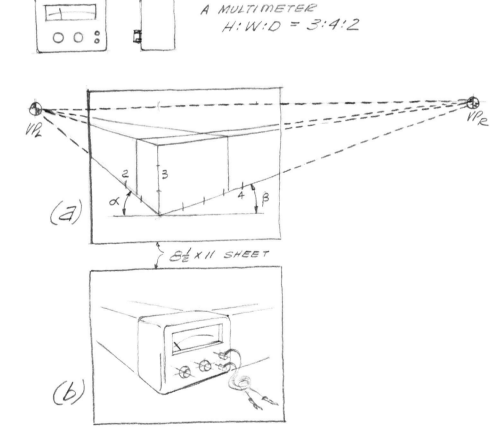

We know that we do not want to set α and β such that $\theta < 90°$. However, α and β are variables and the relation between them determines how we view an object in the following way:

If $\alpha > \beta$, the right face of the object is exposed to view more than the left face.

If $\alpha < \beta$, the left face is exposed more than the right face.

If $\alpha = \beta$, both left and right faces are exposed equally (desirable when depicting long, thin objects).

The solid-line construction of Figure 40a shows the first step in setting up a perspective sketch of a multimeter on $8\frac{1}{2}$ by 11 inch paper. The dashed lines indicate what is estimated, but not drawn. The meter is to be drawn with emphasis on the right face ($\alpha > \beta$). Once this amount of construction has been established, the steps of blocking-in, adding detail, and finishing can proceed by merely observing the rules of convergence to the *VP*'s and foreshortening. Figure 40b shows the sketch up to the finishing stage.

The intent of the above discussion has been to present perspective sketching as a useful and effective tool in the communication of engineering information. A great feeling of accomplishment can come from a quickly made idea-sketch that makes the idea look as it should look.

chapter

9

TECHNIQUES OF SKETCHING

Sketch lightly and quickly with a soft pencil. Embodied in this simple statement lies much of the success and satisfaction of using graphics in engineering. Making drawings for any purpose is a time-consuming task. A fast and well-executed freehand sketch can carry as much or more information than a carefully prepared instrument drawing. Much can be learned about technique from the artist even though the engineer need not have the proficiency of the artist in order to be able to rely on sketching as an important tool of communication. Here are some hints that should help develop sketching ability:

1. *Sketch lightly* for construction lines. Bear down for finished lines. Let a single, soft pencil (HB or softer) do the work.
2. *Sketch freely.* Don't worry about extraneous construction lines if they are light. Feel for a line or a shape rather than expect to draw the finished product on the first attempt.
3. *Use the eraser sparingly.* If your construction is light, erroneous lines will not detract. Extensive erasing and revising wastes time.
4. *To sketch a straight line between two known points,* start with the pencil on one point. Look at the second point as you draw—not at the line as it progresses. Try a quick, short trial line in the direction of the second point. If it appears to be correct, proceed with the line; if it is off direction, make a second attempt, compensating for the original error.
5. *To draw a line parallel to a given line,* position the pencil at the desired starting point and gaze beyond the plane of the paper (as in a vacant stare) as you draw. You will see both lines rather than focusing your gaze on just one line or the other. This is especially useful in establishing horizontal and vertical lines on a sheet of paper. The given lines are the top, bottom or side edges of the sheet.
6. *To make major revisions in a sketch,* trace over the original. Often a sketch that is the result of the development of ideas becomes com-

plicated and involved. Maybe the process of developing the idea has revealed some major flaws that would involve much erasing and revising. Make a fresh start by tracing, on translucent paper, just the ideas and constructions that you wish to retain. This new drawing can then be used for further development until it, too, must be traced.

Artists and illustrators use this technique freely—often proceeding through dozens of revisions and fresh starts before arriving at one that suits them. They use pads of paper designed for this purpose. It can be obtained at art supply stores under such names as Visualizing paper or layout tracing paper. The paper is less expensive than engineer's vellum and has better surface quality for freehand work.

7. *To transfer a finished sketch onto opaque paper,* use a modified carbon-paper technique. At times, a rough idea sketch must be prepared for a formal presentation. The tracing technique is ideal for this unless the finished drawing must be on opaque paper. Turn the sketch over and scribble over the entire area of the drawing with a soft pencil. Remove excess graphite, position the sketch, right side up, on the clean sheet and carefully trace over the portions to be transferred. Use a sharp, hard pencil (4H) for this. This will leave an erasable, graphite image on the opaque sheet. The new drawing can then be completed for presentation as desired. Do not use standard typewriter carbon paper. It leaves a greasy image that cannot be erased or revised further.

8. *Line shading* can be used effectively to help develop form in a perspective sketch. Consider the cylinders of Figure 41. At 41a, the outline drawing of a right circular cylinder is shown. It is correctly drawn and one can recognize the shape. At Figure 41b, the same cylinder has been shaded to bring out the roundness of the shape. Simple, bold lines have been used to shade the object. A rounded object will have a gradation of tone from dark to light. This is achieved

(a)

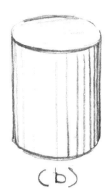
(b)

Figure 41. A right circular cylinder showing the use of line shading to emphasize the curved form.

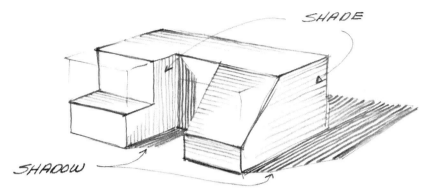

Figure 42. Shading a rectangular object. Shade lines are evenly spaced and are used to show a contrast between adjacent planes and between the object and its background.

in line shading by varying the spacing between the shade lines (varying the percentages of black and white.)

Figure 42 shows a rectangular object with shading. Here the shade lines are kept evenly spaced so as to give the impression of a flat plane and not a rounded surface. In shading engineering drawings, the light source is usually considered as coming over the left shoulder toward an object in front of the observer. Surfaces on the opposite side of an object from the light source are considered to be in *shade*. Surfaces on the object that receive no light because of the intervention of another object or part of the same object are said to be in *shadow*. There is a definite geometry of projected shade and shadow.

An artist may spend more time in shading (or rendering) a drawing than he will in developing the basic form. He can show material, texture, lighting conditions, reflections, contour of the supporting surface, etc. through shading. For the engineer, shading should be used as an aid to clarity! It should be done quickly for the purpose of reducing any ambiguity that might confuse the observer.

9. *Practice*. Success at developing any skill depends upon the amount of practice one is willing to do—and freehand sketching is a skill. In spare moments, sketch what is in front of you. Work quickly and freely. Don't strive for a final, realistic rendering. Try to find shapes and draw them. Draw your own hand—an interesting and challenging subject. It is made up of a series of cylinders and prisms. The problems of proportion and foreshortening are difficult if you are to get realistic results.

chapter

10

CHARTS, GRAPHS, AND MATHEMATIC

CONSTRUCTIONS

Graphics, as a working language of the engineer, has been presented so far as a tool for depicting physical objects. There are many other forms of drawing which can well serve the engineer in all phases of the design process. These are charts, graphs, and mathematic constructions—tools of analysis and presentation.

The business and technical worlds are fast becoming worlds of statistics. Modern methods of machine record storage and retrieval make it possible to gather continuous data on any type of business or industrial operation. The techniques and procedures of systems engineering and operations research require detailed data on which to base immediate decisions and predictions for the future. Decisions on industrial management problems which were previously left to human judgment (often guesswork) or ignored as being too complex are now being analyzed statistically according to mathematic rule. One result of this emphasis on statistical decision making is that, as happened to the Sorceror's Apprentice, we are becoming inundated with enormous masses of data that must be analyzed and used. And, more importantly, the analyses must be presented to and understood by many individuals of widely differing backgrounds. Charts and graphs have proven to be the clearest, most concise way to present numerical data. For the design engineer, familiarity with the use of charts, graphs, and mathematic constructions will provide good working tools for analyzing and, ultimately, presenting his ideas.

Definitions of these three drawing forms are as follows:

Charts. The display of data that does not necessarily follow mathematical rule.

Graphs. A device for analysis and investigation in which relationships or laws involved in numerical data are represented by means of curves or other figures.

Mathematic constructions. The process of solving problems by graphic constructions which follow geometric rule and result in a numerical or pictorial answer.

Discussions of the details of these three drawing forms are either too simple or too complex for this text. The simple forms should be obvious. Their uses are limited only by the user's imagination and ability to organize. The complex forms follow rigid rules which are beyond the scope of this discussion. The following is a brief description of the most useful forms. More detailed descriptions may be found in engineering drawing and graphic science texts.

CHARTS

Chart forms useful to the design engineer are:
1. Block diagrams.
2. Flow charts and process charts.
3. A selection of forms to show the fluctuation of numerical variables, i.e., multiple bar charts, pie diagrams, 100 per cent bar charts, etc.

Block Diagrams

The block diagram is becoming a standard symbol of scientific and engineering work. An engineer will invariably construct a block diagram when making a presentation to others, whether he is talking about a Venus-probe missile system or a device to measure noise in roller bearings. He quickly sketches and labels a series of rectangles, triangles, circles, ellipses, or an occasional fried-egg shape (if the concept of that block is not clear) and interconnects them in a myriad of ways with lines and arrowheads. This is a block diagram. It gives an overall view of a system of thoughts, devices, processes, or concepts and shows, in a qualitative manner, how these elements may be interrelated. His immediate task may involve only one small element of the diagram, but a knowledge of the entire system is essential to the complete understanding of the parts.

A block diagram is easy to draw. There are no rules or standards to follow. Little time should be spent in its preparation. Rectilinearly ruled paper is an aid in preparing block diagrams. Figure 43 shows the design process in block-diagram form. All of the elements are shown, their interrelations are pointed out, and the iterations of the process are suggested.

A block diagram cannot show the intimate details of the system it represents. These are only discovered by the designer himself as he proceeds

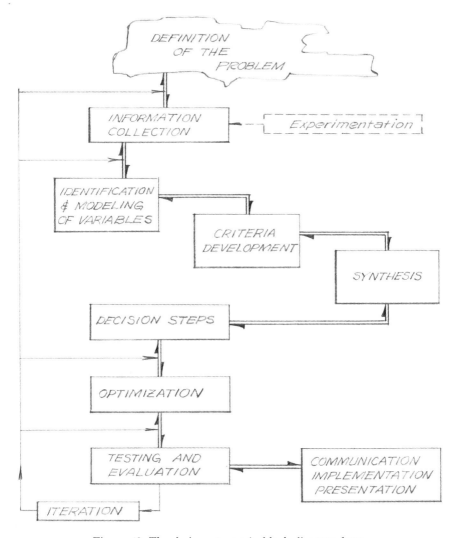

Figure 43. The design process in block diagram form.

toward a solution. The uninitiated, however, can gain a better understanding of the designer's problems and goals if his presentations are supported by block diagrams.

Flow Charts

The flow chart is similar to the block diagram but is used mainly to show the steps in an established process. This process could be a chemical

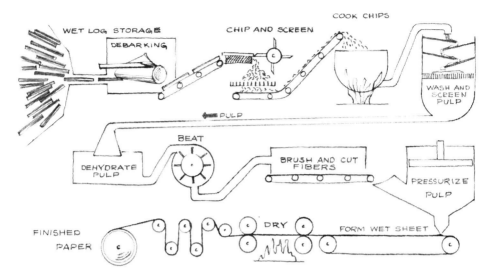

Figure 44. A flow chart of a paper-making process. This chart was designed to show the organization of the principal steps in the process from raw material to finished product.

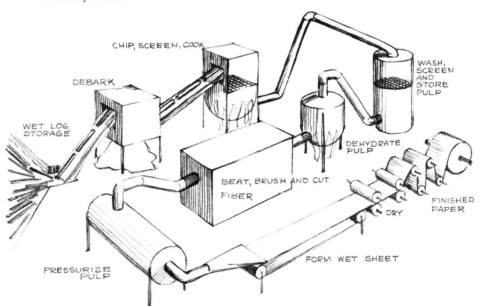

Figure 45. The same paper-making process shown as a perspective sketch. Note that simple, bold line shading has been used to bring out the three-dimensional shapes.

process, a manufacturing production process, a product distribution system, or simply the flow of paper work through a large company. They differ from block diagrams in that they often show the steps in the process pictorially. In a chemical process, for instance, the process may start with a pile of raw material, illustrate schematically such operations as cleaning, grinding, blending, baking, molding, and packaging, and finish with the product leaving the shipping dock. Figure 44 shows such a flow chart. The schematic pictures add a tone of realism to the chart. The principles of perspective can well be used in flow charts to heighten this sense of realism. Figure 45 is the same process drawn in perspective.

Other Charts

The statistics that surround us often include variables whose fluctuations do not follow a mathematical law. These fluctuations, however, are important in analyzing performance, making predictions, or demonstrating progress (or failure). These variables could include the number of automobiles sold in southwestern United States in 1962, the mean rainfall in Cedar Rapids, Iowa, since 1900, or the number of freshmen entering college for the past ten years. In general, the information is more popular than technical. Because of this, effort must be made to present it in a simple and easy to grasp form. For this, the bar chart, the pie diagram, and variations of these forms are used. Figure 46 shows a few examples of these kinds of charts. The manner in which they are presented is limited only by the user's imagination. Business and news magazines, and the newspapers use this form of chart to get data across to their readers quickly and easily. They have developed techniques of presentation that are worthy of study by anyone using this form. The use of color for coding and textural changes in black and white are notable examples.

GRAPHS AND MATHEMATIC CONSTRUCTIONS

The technique of analysis by graphic methods is a useful tool for the design engineer. By definition, the graphic method is "the method of scientific analysis and investigation in which relations or laws involved in numerical (or geometric) data are represented to the eye by means of curves or other figures." With these techniques, the design engineer can display data for study and analysis and perform certain mathematic manipulations on the data.

The design process is filled with decision-making steps for which we must first have information gained by analysis or experiment before we

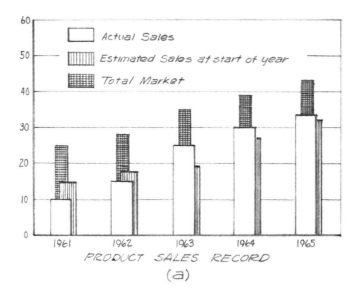

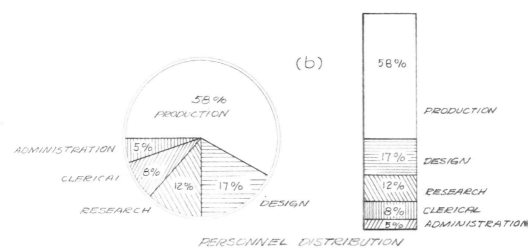

Figure 46. Popular charts. The bar chart at (a) shows the reader at a glance that 1. the five-year growth of product sales has been steady and substantial, 2. the company has captured a large proportion of the market, 3. they have improved on their forecasting ability, and 4. the total market is leveling off. Two forms of 100% charts are shown at (b), the pie diagram and the 100% bar. These give a quick visual interpretation of the percentage figures.

can proceed. Often the information causes us to go back and re-examine earlier decisions. The graphic method is ideal for a quick examination of progress at an early stage of the process and is accurate enough to base early decisions on before the need for more precise analysis.

As a design proceeds into the experimental stage, empirical data is produced. The engineer will instinctively plot this data (if it is in numerical form). Many kinds of graph paper are readily available and most engineers learn early to have it handy and use it. Graphic methods involve manipulating this data to obtain further information. These methods may take the following forms:

1. Determination of empirical equations
2. Graphical integration and differentiation
3. Construction of nomograms
4. Solution of vector problems

All of these areas of graphic mathematics are too complex for a detailed discussion here. However, their uses and usefulness in engineering design should be mentioned. It should also be pointed out that the graphic method is only approximate—the degree of accuracy being dependent upon the precision of drawing. Information upon which to base final design decisions should come from rigorous analytic treatment or from computer analysis.

Empirical Equations

Experimental work often yields data on the relation of such variables as temperature and time, velocity and time, voltage and current, etc. As was stated before, the engineer usually plots this on ruled paper, it being easier to "read" the data in graph form than in tabular form. If the data plots in a straight line (or approximately straight line) that line can be drawn and its equation determined from the general expression $y = mx + b$, where m is the slope of the line $\Delta y / \Delta x$ measured in units of the variables x and y, and b is the intercept on the y axis when $x = 0$.

If the plot is not a straight line, the curve may represent either the power equation, $y = bx^m$, or the exponential equation, $y = bm^x$. These two forms are common in nature, so that experimental data could easily follow one or the other. When plotted on rectilinear paper, the resulting curves would fall into categories as shown in Figure 47a and b. Taking the logarithms of these two equations yields:

For the Power equation $\quad \log y = \log b + m \log x$
For the Exponential equation $\quad \log y = \log b + x \log m$

Examination of these reveals that they are both of the form $y = mx + b$ if the logarithms of both x and y are plotted in the first case and if $\log y$

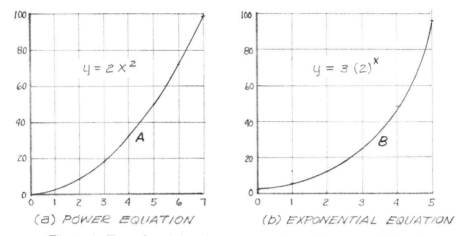

Figure 47. Examples of (a) the power equation of form $y = bx^m$ and (b) the exponential equation of form $y = b(m)^x$—two forms occurring frequently in nature. Empirical data that plots in one of these forms can be rectified as shown in Figure 48.

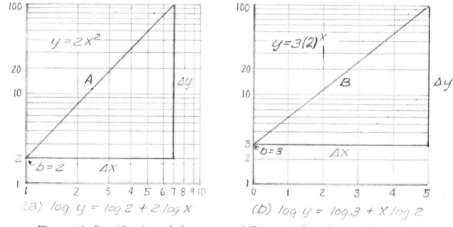

Figure 48. Rectification of the curves of Figure 47 by plotting the data on log-log paper at (a) and on semi-log paper at (b). From these straight-line plots, the equations can be found by using the general form $y = mx + b$.

and x are plotted in the second. These should yield straight lines from which the equations can be written as before. Looking up the logarithms for a mass of data is a tedious job, but, fortunately, this is not necessary since logarithmic graph paper is available. It is only necessary to plot the value of the variables x and y on the correct paper to obtain the straight line. Figure 48a shows the transformation of curve A (Figure 47a) when plotted on *log-log paper* and Figure 48b shows the same transformation of curve B (Figure 47b) when plotted on *semi-log paper*. From these two straight-line plots, the respective equations can be determined.

Graphic Calculus

Empirical data of displacement versus time was obtained by experiment and plotted on rectilinear coordinates to yield the curve of Figure 49a. It is known that the first derivative of this data will yield velocity versus time. The displacement equation may not be known, but by graphic methods, the velocity curve can be determined. Figure 49b shows this curve. Similarly, the second derivative of displacement will give acceleration versus time. Figure 49c shows the acceleration curve obtained by

Figure 49. Graphic calculus. An empirical time-displacement curve (a) is differentiated to obtain a time velocity curve (b). The second differential yields the time-acceleration curve (c). Graphic integration of (c) would yield (b) and integration of (b) gives (a) since integration and differentiation are reversible processes.

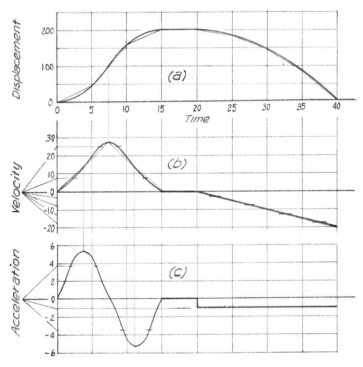

differentiating the velocity curve graphically. Thus, with only the empirical displacement curve given and no knowledge of its equation, a complete picture of the velocity and acceleration pattern can be obtained graphically with a good degree of precision.

If the original data yielded velocity versus time, the integral curve of this would yield displacement versus time. Graphic integration and differentiation are reversible processes. The techniques are simple. Any good graphics text gives a detailed description of the theory and procedures involved in graphic calculus.

Nomography

A nomogram is a chart by which the solution of an equation relating three or more variables can be easily found over a wide range of the variables. It is a graphic form most useful if repeated calculations of the equation must be made. Figure 50 shows a nomogram for the equation

Figure 50. A nomogram for the equation $0.1X + 0.75Y = Z$. Any straight line drawn across the figure (such as dashed line A) will give a solution to the equation. Relatively easy to construct, a nomogram is most useful when many solutions of an equation must be made.

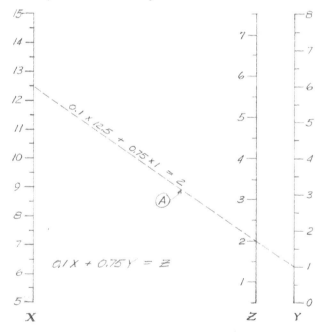

$0.1x + 0.75y = z$ with x ranging from 5 to 15 and y ranging from 0 to 8. Choosing any values of x and y and drawing a straight line between these two values on the appropriate scales will yield the value of z that satisfies the equation.

Nomograms are not difficult to construct, but they must be drawn and calibrated carefully to maintain precision.

Vector Geometry

A vector quantity is one having magnitude, direction, and position. These three qualities allow us to express a vector quantity graphically as a straight line. The magnitude is the length of the line drawn to an appropriate vector scale; the direction is denoted by an arrowhead at one end, and the position is the orientation of the line as dictated by the specific problem. With the understanding of a few basic principles, we can add and subtract vectors and obtain the resultant of two or more vec-

Figure 51. These forces, F_1, F_2, and F_3, operate on a body. The forces are resolved graphically to find the magnitude, direction, and position of the resultant of the three forces.

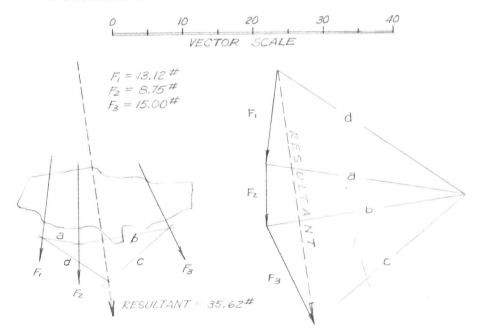

tors graphically. The solution of static force problems in two or three dimensions can be found by graphic means. Analyses of dynamic force or velocity systems can be made for instantaneous situations. This is especially useful in kinematic design. Figure 51 shows the resolution of three coplanar forces by the string polygon method. The resultant is shown in magnitude, direction, and position. The techniques of vector geometry are simple. Drawing precision must be maintained since numerical answers are derived from scale measurements.

Descriptive Geometry

Descriptive geometry is a subject concerned with the precise graphic description of points, lines, planes, and curved surfaces in space using orthographic projection techniques. Problems involving the relations between lines and planes in space, the intersection of planes and solids, or

Figure 52. Descriptive geometry—the use of orthographic projection techniques to make precise solutions of space problems. Here the line of intersection of a plane (*ABC*) and a cone are found using the basic concepts of true length of a line, point view of a line, edge view of a plane, and true shape view of a plane to locate and plot the true shape of the line of intersection.

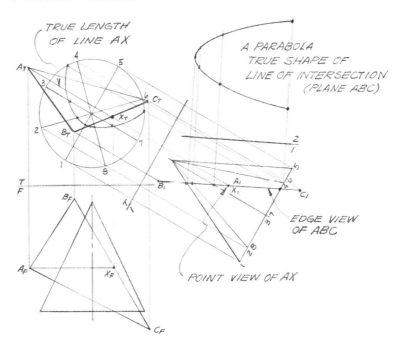

the intersection of two or more solids would have to be studied by building three-dimensional scale models if no accurate means were available to represent complex spatial relations on a two-dimensional drawing surface. The orthographic projection system is ideally suited for use in solving such problems.

An infinite number of orthographic views of an object or arrangement of objects can be projected once two views (usually top and front views) have been defined. A successful solution is obtained by knowing which auxiliary views to take and by following a few simple rules. Figure 52 shows the orthographic construction required to find the line of intersection between a plane and a cone.

The basic operations of descriptive geometry are finding the true length and point views of a line and finding edge and true views of planes. Detailed discussions of these constructions are not within the scope of this text. Many good descriptive geometry texts are available showing applications of basic operations to a wide variety of problems.

Freehand Mathematic Constructions

Most of the work in mathematic constructions should be done carefully with instruments. However, a good understanding of basic graphic theory and a skill in freehand sketching—especially in sketching straight and parallel lines and proportioning distances—permits the use of freehand techniques for the approximate solution of some problems. Approximate quantitative results are often good enough to prove or disprove original assumptions or to provide information needed for a more accurate, final study.

If graph paper is not available, data can be plotted quickly using sketching techniques and proportioning the scale calibrations by eye. With practice, freehand graphs can be made to a fair degree of accuracy —usually good enough to make a quick examination of the data at hand. Logarithmic scales can be calibrated approximately by estimating the principal logarithm-cycle divisions from the knowledge that $\log 1 = 0$, $\log 2 = 0.30$, $\log 3 = 0.48$, $\log 4 = 0.60$, $\log 5 = 0.70$, $\log 6 = 0.78$, $\log 7 = 0.85$, $\log 8 = 0.90$, $\log 9 = 0.95$, and $\log 10 = 1.0$.

chapter

11

PRESENTATION

Chapters 7, 8, 9, and 10 have dealt mainly with the theory and techniques of drawing with emphasis on the communication of the engineer with himself. A successful designer will be increasingly called upon to present his work to others. Some of the persons or groups of persons interested in his work could be:

1. Company management
2. Shop personnel
3. Sales personnel
4. Other companys' salesmen supplying components
5. The design engineer's colleagues
6. Armed force liaison people and other government agents
7. The general public
8. Visiting school children
9. Members of technical societies
10. Industrial designers, advertising men, package designers, etc.

People are interested in knowing more about new ideas. The spectrum of knowledge of technical matters of this wide audience can run the gamut from zero to considerably more than the author himself. In making a presentation, whether oral, written, graphic, or any combination of the three, the material must be carefully tailored to the audience if maximum information is to be transmitted with minimum noise.

Graphic presentations useful to the design engineer fall into three categories: report illustrations, slides, and posters.

REPORT ILLUSTRATIONS

Two important questions to answer in planning illustrations for a written report are what illustrations will best enhance the transmission of information by means of the report? and what reproduction process is to be used to duplicate the report? The first question should be posed at the time the first outline is prepared. Too often illustrations are an afterthought—added after the writing is finished with more intent of breaking up the monotony of the typed page than enhancing the text. A good illustration can eliminate a lot of descriptive text. This is a most important consideration for these times when competition for reader interest is fierce. Well-drawn illustrations with complete and concise captions can allow a reader to get a good impression of the work reported by concentrating his attention on figures alone. The popularity of picture magazines attests to this fact.

Of course, all technical reports do not lend themselves to illustrative treatment. The important point here is that often no consideration is given to the help that some good illustrations can give the readers of technical reports. Engineering, by its nature, deals more with physical objects and numbers than with abstract thought. The chances are excellent that the results of engineering effort will have many concepts that can be presented graphically.

The answer to the second question posed above dictates how an illustration is to be prepared. The following are the most common duplicating processes used in business, industry, government, and education, with comments on the problems of preparing illustrations for each:

Hectograph (spirit duplicators). Drawings are made in pencil directly on the master. Fine details and variation in line weights are difficult to obtain.

Mimeograph (stencil duplicators). A wax coated, fine-mesh stencil must be cut with a stylus. Drawing is difficult and it takes much practice to obtain good results. Process produces fuzzy images on soft paper. New methods are available to make machine-cut stencils by electronic scanning of inked originals.

Offset printing (office-type machines). Drawings are made directly on a paper master using a special pencil. The process gives a faithful reproduction of the original. Drawing is not easy. Care must be taken not to smudge or leave fingerprints on the master. Photo-offset method uses a photographic negative of inked original drawings in the preparation of the master. Reproduction is excellent and the original may be enlarged or reduced. The cost of the negative must be considered.

Many installations now use the Xerox method of preparing masters. They are made optically from original art work without the costly negative. The results are good but not quite as sharp as the photographic method. The process does a good job of reproducing pencil originals, thus saving inking time. Originals may be enlarged or reduced.

Offset Printing (commercial). Commercial printers prefer inked originals. Reproduction is excellent. Originals can be enlarged or reduced.

Diazo Printing (Blueprint or blue line). This process is economical up to a maximum of 20 copies of relatively short reports. Typing and art work are done on translucent vellum. Orange-colored carbon paper is available to use as a back-up for typing and drawing. It puts a reverse image on the back of the sheet, increasing the density of the line for sharper prints. Diazo prints are not permanent and will fade with time.

Some other considerations in preparing effective report illustration are lettering, scale, and the use of photographs.

Lettering

Well-executed hand lettering in engineering block letters is the fastest and least expensive way to add notes and numerals to a report drawing. However, unless the letterer has considerable skill and smooth, even style, hand lettering is best avoided. Lettering guides such as the Leroy and Wrico guides are the best solution to the problem of lettering. They produce uniform, sharp letters but some practice is needed to achieve even letter spacing. Where the budget permits or copy is limited to a few words, various kinds of paste-on letters can be used. These are excellent reproductions of type-set letters and come in a range of sizes. They are printed on an acetate or clear paper sheet and can be lifted off individually and composed into words directly on the drawing.

Scale

For a given report, it is best to have all illustrations drawn to the same scale, whether it is full scale or larger or smaller than full size. If drawings are made to different scales, such things as lettering size and line width will vary from drawing to drawing. The only way to avoid this problem is to take care in proportioning letter size and line width to the particular enlargement or reduction planned for each drawing.

Photographs

Photography, though not related directly to graphics, plays an important role in report illustration. A good photograph is the best means for showing results. If the reader sees a photograph of experimental equipment or a finished device, he knows that the work was accomplished. A drawing could represent the author's plans of what he hopes to accomplish.

Good photographs are more expensive than drawings to both prepare and print. If you are not adept at photography, hire an expert to take your pictures. The success of a black-and-white print depends more upon the darkroom work than on the camera work. Ask for a print with wide contrasts of light and dark. Any printing process will drop out some of the contrast of the original. Weak photographic prints will yield a printed copy from which much of the detail has disappeared. All printing processes require a halftone (or screened) negative for the plate-making operation. Supply the printer with a sharp, glossy print—not the negative.

A weak photograph that is not suitable for printing can often be used as a base for making a sketch by tracing. Use a soft pencil or a fine artist's pen on translucent vellum placed over the photograph. Keep the drawing simple, use line shading and a few solid blacks for contrast, and emphasis.

SLIDES

An oral, technical report is greatly enhanced by the use of a few well-designed slides. Slides are becoming increasingly popular with speakers, but nowhere is there more abuse of the concepts of effective communication. Some common faults in the preparation and use of slides are:

1. The speaker has photographed illustrations from his paper or a textbook for his slide material. Usually these sources contain far too much information for a slide and are not designed for projection. Often lettering and line work cannot be read farther back than the first few rows of the audience.
2. The speaker has a list of 14 points that he wishes to make, so he types 14 sentences on a piece of paper and photographs it. Few people can read it and, if they can make out the words, they concentrate so much on this task that they don't listen to the speaker. Typewriter copy should be used only when there are four or five words to be presented and the ratio of film size to original size is close to 1:1.
3. A speaker has a 30-minute talk and presents 30 to 40 slides. They go by so fast that the audience either misses the slides or the speech.

Some points to keep in mind in preparing slides are:

1. Keep copy to a minimum and keep art work simple. Use bold lines and large lettering.
2. Use color whenever possible for coding, emphasis, or decoration.
3. Don't fill the projection area. Keep one-third of the area blank.

4. Choose slide material carefully. Don't have too many slides. One slide for every three to five minutes of the talk is ample.
5. Let the slide create an interest in what *you* are saying—do not let the slides say it for you. The audience will not be listening if they must study a slide.
6. The slide should not repeat exactly what you have to say, but should say it in another way. Slides should be used to introduce redundancy into your presentation. (Observe television commercials. The words on the screen state the message simply while the voice delivers the same message in more detail.)

Slide Systems

The most common types of slide and projector systems in use today are:

1. *2 in. by 2 in. (35 mm)*. This is the most inexpensive system for obtaining full-color transparencies. Full use of this advantage, however, calls for the production of colored originals.
2. *$3\frac{1}{4}$ in. by 4 in.* The common $3\frac{1}{4}$ by 4 glass slides are going out of vogue with the advent of the Polaroid positive transparency film. Glass slides in color are quite expensive. Black and white slides of any type of original material can be made quickly, easily, and inexpensively using the Polaroid system.
3. *8 in. by 10 in. Transparencies for Overhead Projection.* New materials are making the overhead projection system increasingly popular. There are many distinct advantages to this system.
   a. The speaker has control of the projector and the slides.
   b. Room lights can be left on and window shades kept up unless the room is unusually bright.
   c. As many as five or six overlays can be progressively laid on the original to build a point or a problem solution.
   d. The transparencies are inexpensive and are easy to make without the need for costly photographic equipment.
   e. Extensive color can be used without the necessity for colored originals. All originals are black on white.
   f. The speaker can add to the slide by writing or drawing directly on it while it is being projected.
   g. During question periods, the speaker can easily go back to an early slide to strengthen a point without awkward communication with a projectionist.

Slide Preparation

The following are some suggestions for preparing copy or illustrations to be made into slides:

1. *Colored originals.* If the color work is to be complicated and extensive, hire an artist to prepare the original. A simple color treatment is often just as effective as an ambitious one. Curves can be laid down using colored tapes that come in various widths. Some materials for use in making originals are colored ink, transparent and opaque water colors, colored pencils, felt pens and brushes, and colored paper.
2. *Size.* Always prepare originals to the same proportion as the frame size of the film. For 2 in. by 2 in. (35 mm) and $3\frac{1}{4} \times 4$ in. slides, the ratio of height to width is $1:1\frac{1}{2}$. Do not crowd the edge of the frame with copy. Leave a generous border on all four sides.
3. *Polaroid Slides.* Polaroid transparency film is panchromatic, so practically any original is suitable for slide work. An interesting effect can be gained by preparing the original on light grey paper using white and black for the drawing. Use white conte pencil, wax pencils, or opaque water color (Chinese white) for the white copy and soft pencil or india ink for the black.

POSTERS

Slides are ideal as a visual aid for a large audience in a large room. Often, however, engineering ideas are to be presented to a few people in a small conference room. Here posters can be useful. Available poster materials vary, but, in general, illustration board in 20 in. by 30 in., or 30 in. by 40 in. size is ideal. Illustration board is heavy enough to stand by itself on an easel or a shelf. The drawing surface is good and will take pencil, ink, water color, tempera paint, etc., readily.

Rules for preparing posters are similar to those for preparing slides. Make the drawings and layout simple. Use large, bold lettering and keep copy to a minimum. Letters should be at least one-inch high. Tailor the lettering to the size of the room to be used for presentation. Use color for emphasis, coding, and decoration.

Flip charts are popular, both for prepared aids and on-the-spot presentations. An easel and a pad of newsprint (about 30 in. by 40 in.) are the required equipment. Use heavy crayons, chalk, or felt pens for writing and drawing. If you wish to "turn off" the visual presentation for a moment, interleave blank sheets so the old display can be flipped out of the way and the new will not be exposed until you are ready for it.

Many engineering offices are being equipped with chalkboards. The chalkboard is an ideal aid to small-group communications. Participants can gather closely around a board. Each person can see and contribute easily. Creative design problems often require an incubation period before a workable solution emerges. Ideas put on a chalkboard are constantly in front of you and can be observed and revised over a period of time. Photographs can be taken of ideas worked out on a chalkboard for record purposes. Cork bulletin boards have the same advantage as far as posting ideas for continued study and evaluation.

SUMMARY

The last five chapters have been an account of the language of engineering graphics—what the language is and how it can be used by the design engineer. As with any language, it is a system of codes, the symbols being lines and the coding system being the manner in which the lines are combined to form images capable of transmitting information. To use the system effectively, the designer must understand the theory of graphics and develop skill in drawing through practice.

Nothing is more satisfying than having another person—the receiver—ask, with a note of admiration in his voice, "Did you draw that? It's good!" Chances are that information has been transmitted successfully.

BIBLIOGRAPHY

Engineering Graphics

French, T. E., and C. J. Vierck, *Graphic Science* (2nd ed.). New York: McGraw-Hill Book Company, 1963.

Levens, A. S., *Graphics*. New York: John Wiley & Sons, Inc., 1962.

Luzzader, W. J., *Basic Graphics*. Englewood Cliffs, N. J.: Prentice-Hall, Inc., 1962.

Rule, J. T., and S. A. Coons, *Graphics*. New York: McGraw-Hill Book Company, 1962.

The above four texts each give excellent coverage to all aspects of engineering graphics. They are written for the engineering student and not the drafting trainee.

Other Texts

Doblin, J., *Perspective*. New York: Franklin Watts, Inc., 1956. A unique treatment of perspective drawing by an industrial designer for anyone interested in learning to draw or sketch quickly and accurately in the perspective form.

French, T. E., and C. J. Vierck, *Engineering Drawing* (9th ed.). New York: McGraw-Hill Book Company, 1960. A standard in the fields of engineering drawing and drafting practices since 1911.

Katz, H. H., *Technical Sketching and Visualization for Engineers*. New York: The Macmillan Co., 1949.

Thomas, T. A., *Technical Illustration*. New York: McGraw-Hill Book Company, 1960.

Two volumes that present the techniques of freehand technical sketching.

Guptill, A., *Pencil Drawing Step by Step*. New York: Reinhold Publishing Corp., 1959. A standard work on pencil sketching by an artist.